Compound Systems in Nonextensive Statistical Mechanics and Manganites

Compound Systems in Nonextensive Statistical Mechanics and Manganites

By

Felipe Américo Reyes Navarro

Cambridge
Scholars
Publishing

Compound Systems in Nonextensive Statistical Mechanics
and Manganites

By Felipe Américo Reyes Navarro

This book first published 2022

Cambridge Scholars Publishing

Lady Stephenson Library, Newcastle upon Tyne, NE6 2PA, UK

British Library Cataloguing in Publication Data
A catalogue record for this book is available from the British Library

ISBN (10): 1-5275-8445-3
ISBN (13): 978-1-5275-8445-7

To

my parents Américo and Julia,
and my brother Gustavo

Everything is theoretically impossible, until it is done.
—Robert A. Heinlein

If I have seen further, it is by standing on the shoulders of Giants.
—Isaac Newton

Impossible only means that you have not found the solution yet.
—Anonymous

The measure of intelligence is the ability to change.
—Albert Einstein

The snake which cannot cast its skin has to die. As well the minds which are prevented from changing their opinions; they cease to be mind.
—Friedrich Nietzsche

All I need is a sheet of paper and something to write with, and then I can turn the world upside down.
—Stephen Hawking

TABLE OF CONTENTS

List of Figures xi

List of Tables xii

Acronyms xiii

Foreword xiv

Preface xv

Introduction 1

Part I Manganites

Chapter One 4
Introduction to the Physics of Manganites
A Brief Historical Outlook 4
General Formula 5
Crystal Structure 5
Pyrochlore Manganites 7
Electronic Structure 8
Main Theories Seeking to Unravel the Physics of Manganites 8
Phase Separation Scenario 12
Charge-Ordering Phase Transition 14
Orbital Ordering 15
The $Pr_{1-x}Ca_xMnO_3$ Series 15
A Summary of Some Models for Manganites From First Principles .. 16
Nanostructured Manganites 17
Potential Technological Applications 18

Part II Nonextensive Statistical Mechanics

Chapter Two 22
Nonextensive Statistical Mechanics
Historical Events 22
Entropies Successfully Applied in Different Areas of Science 28
Thermodynamic Entropy 28
Boltzmann Entropy 28
Shannon Entropy 29
Quantum Group Entropy 29
κ–Entropy 30
Superstatistics 30
Basic Concepts of the Third Version of Nonextensive Statistical Mechanics 31
Tsallis Entropy 32
Demonstrating How to Obtain Tsallis Entropy Without Postulating It 33
Demonstrating that the q-Exponential and q-Logarithm are Inverse Functions 36
The Connection Between Renyi Entropy and Tsallis Entropy .. 37
Pseudo-additivity for Tsallis Entropy 38
Definition of Temperature 39
Nonextensive Quantum Mean Values 40
Escort Distributions 42
Escort Entropy 43
Entropy Optimization: Maximization and Minimization 44
Some Demonstrations by Using the Maximum Entropy Method 45
Boltzmann-Gibbs-Shannon Entropy Maximization 45
Boltzmann-Gibbs-Shannon Entropy Maximization for a Canonical ensemble 45
Boltzmann-Gibbs-Shannon Entropy Maximization for a Grandcanonical Ensemble 46
Tsallis Entropy Maximization in the First Version of Nonextensive Statistical Mechanics 48
Tsallis Entropy Maximization in the Second Version of Nonextensive Statistical Mechanics 50
Tsallis Entropy Maximization in the Third Version of Nonextensive Statistical Mechanics 51
Demonstrating What the Lagrange Multipliers Are (in the BGS Entropy Maximization) 54

Demonstrating What the Lagrange Multiplier α Is (for a Canonical Ensemble) 55
Demonstrating What the Lagrange Multiplier β Is (for a Canonical Ensemble) 55
Demonstrating What the Lagrange Multiplier α Is (for a Grandcanonical Ensemble) 57
Demonstrating What the Lagrange Multipliers β And γ Are (for a Grandcanonical Ensemble) 57
Demonstrating How to Obtain Boltzmann Entropy from Boltzmann-Gibbs-Shannon Entropy 59
Demonstrations of Several Properties of the q-Exponential Function ... 60
Finding Out a Formula for the q-Exponential Function of the Sum of Two Functions 60
Finding Out a Formula for the q-Exponential Function of the Sum of Three Functions 61
Demonstrating a Formula for the q-Exponential of the Sum of n Functions 62
Finding Out a Formula for the Product of Two q-Exponentials 64
Finding Out a Formula for the Product of Three q-Exponentials 65
Finding Out a Formula for the q-Exponential of the Difference of Two Functions 66
Finding Out a Formula for the Division of Two q-Exponentials .. 67

Chapter Three 69
An Alternative Method to Calculate Mean Values of Compound Systems
A Formal Derivation from the Full Hilbert Space Toward the Hilbert Subspaces 69
On the Exemplifying Magnetic Model to Be Utilized 73
Mean Values of a Two Spin-½ System (Third Version of Nonextensive Statistical Mechanics) 74
Applying an Alternative Method to Calculate Mean Values of Compound Systems 76
Additional Calculation for a System $A + B$ With Interaction 77
Some Detailed Calculations 79
Analytical Calculus for $q = 2$ 80
The Escort Distributions of Compound Systems in Nonextensive Statistical Mechanics 82

Part III Applications

Chapter Four ... 86
Application of the Matrices *E*: a Model for $Pr_{1-x}Ca_xMnO_3$ with x = 0.30–0.50
Evidence for Nonextensive Magnetism in Manganites ... 86
Experimental Curves ... 87
Quasi-probability Distributions ... 91
Nonextensive Quasi-probability Distributions ... 92
A Model for $Pr_{1-x}Ca_xMnO_3$ (x = 0.30–0.50) Under Weak Magnetic Fields ... 93
Preliminary Formulae ... 93
The Model ... 94

Chapter Five ... 99
In Summary

Appendix A ... 100
Topics on Magnetism

Appendix B ... 104
Demonstration of Several Properties of the *q*-Logarithm Function

Appendix C ... 111
Demonstration of Several Properties of the *q*-Operations and *q*-Gaussian

Bibliography ... 135

Index ... 137

LIST OF FIGURES

1.1. Cubic perovskite structure.
1.2. Energy levels of the Mn^{3+} ion.
1.3. Major theories for manganites.

3.1. For mean values of a system A + B, a layout showing the summarized formal derivation from the full Hilbert space to the Hilbert subspaces.
3.2. Statistical independence in the third version of nonextensive statistical mechanics.
3.3. For spins $S_A = 2$ and $S_B = 3/2$, magnetization vs. temperature ($M_{A(B)}$ vs. T) by using the third version of nonextensive statistical mechanics and matrices E, too.
3.4. In the third version of nonextensive statistical mechanics, magnetization vs. temperature ($M_{A(B)}$ vs. T) for a two spin-½ system.
3.5. Additivities of total energy vs. temperature ($U_{tot.}$ vs. T) and magnetization vs. temperature ($M_{tot.}$ vs. T).
3.6. For escort distributions, a layout showing the summarized formal derivation from the full Hilbert space to the Hilbert subspaces.
3.7. Use of the escort distributions.

4.1. For Pr0.65Ca0.35MnO3, magnetization vs. temperature measured in a CBPF SQUID.
4.2. For Pr0.65Ca0.35MnO3, magnetization vs. external magnetic field measured in a CBPF SQUID.
4.3. Magnetic entropy curve for Pr0.65Ca0.35MnO3.
4.4. Experimental magnetization vs. temperature for Pr0.65Ca0.35MnO3 under low magnetic fields.
4.5. For Pr0.65Ca0.35MnO3, a comparison between experimental and calculated M vs. T.
4.6. For Pr0.70Ca0.30MnO3, computer calculation of magnetization vs. temperature.

LIST OF TABLES

4.1. Data used in the CBPF SQUID in measuring $Pr_{1-x}Ca_xMnO_3$ with x = 0.35 and 0.40.
4.2. For a cubic symmetry, the conventional Stevens operators.
4.3. Angular momentum quantum numbers of $Pr_{0.65}Ca_{0.35}MnO_3$.

ACRONYMS

FHS: Full Hilbert space
HS: Hilbert Subspace
CBPF (Portuguese acronym)**:** Brazilian Center for Research in Physics
SQUID: Superconducting Quantum Interference Device
CMR: Colossal Magnetoresistance
ZFC: Zero Field Cooling
FC: Field Cooling

FOREWORD

The book "Compound Systems in Nonextensive Statistical Mechanics and Manganites", by F. A. R. Navarro is a material that was actually missing in the literature that treats the interesting field of Nonextensive Statistical Mechanics, as first introduced by C. Tsallis. The subject is extremely rich and has, by now, a vast spectrum of works that cover the area. However, the specific application to compound materials—in this case manganites—opens up new possibilities of learning the topic while considering a relevant category of structures. It is non-trivial application.

The introduction is very well presented and sets up the necessary backgrounds for a pleasant and fluent reading. Navarro is able to pick up the truly relevant points and he presents them with the ability of a long experience in thinking about and teaching the material. It is really a very helpful introduction to fix up the pre-requisites to dive into the interesting chapters that follow up.

The next chapters are more specific, but clear and complete, with the ideas, calculations and discussions presented with remarkable explanations. No doubt that beginners and those who wish to enter the field shall benefit a great deal from the material covered by Navarro's book. The chapter with the applications is very up-to-date and may provide tools and inspiration for further research.

The Bibliography is good and allows the reader to easily grasp the state-of-art of the research field. Students, young researchers, and more experienced professional may find in the book material that may enrich seminars, lectures, and thesis work.

I deeply recommend Navarro's book for the high quality of its text, for the outstanding balance of the contents of each chapter and for bringing to light a non-trivial and very appealing framework to apply Tsallis' nonextensive statistical mechanics.

J. A. Helayël-Neto,
Brazilian Center for Research in Physics
(CBPF).
November 2020.

PREFACE

By this book, I want to divulge two matters, manganites and nonextensive statistical mechanics. They have been my research aims for many years in both the Federative Republic of Brazil and the Republic of Peru. In the former, in the early 2000s, I started a Ph.D. degree in Physical Sciences at the Brazilian Center for Research in Physics, which is located in the Marvelous City of Rio de Janeiro. In the latter, nowadays, I am a tenured professor at the National University of San Marcos in the city of Lima. For many years I wished to write a book on physics; however, because of the duties of everyday life, I postponed that longing.

Before writing this book, I had focused on publishing only scientific papers. Certainly, after obtaining a scientific result, a manuscript requires less time and fewer pages than a regular book does. But now, finally, I have spent my idle hours ordering and updating information about the two aforementioned matters.

Indeed, organizing this book has been a wonderful personal experience. I always liked too much link ideas and words. After reading this book, I hope you, dear reader, will be interested in delving into the issues introduced herein.

Finally, my thanks and gratitude to all who encourage me to move forward with my life as a physicist.

Felipe Américo Reyes Navarro,
Doctor in Physical Sciences,
Lima, Republic of Peru,
August 2021.

INTRODUCTION

This book deals with the application of nonextensive statistical mechanics in compound systems. In particular, the focus is on magnetism and manganese oxides, because these last materials have several important features that enable the application of nonextensive statistical mechanics; in physics, manganese oxides are also named manganites. Nonextensive statistical mechanics, also known as Tsallis statistics, can address the complex problem of manganites in a phenomenological way. Thus, in this book, we will mainly study compound magnetic systems A + B. This is motivated because manganites have magnetic manganese presenting two important valence states, Mn^{3+} and Mn^{4+}, that is, forming two magnetic sublattices.

Besides, we will learn nonextensive quasi-probability distributions, which originate from the relaxation of the Tsallis cutoff. These distributions describe well the fall of the magnetization of the $Pr_{1-x}Ca_xMnO_3$ manganite series, i.e., the metastable metal-ferromagnetic phase (at low temperatures with concentration x = 0.30–0.50). Similarly, an interesting problem and its solution will be introduced; it is concerning the calculation of mean values of compound systems, in different Hilbert spaces.

Currently, there is rich and varied experimental evidence, as well as excellent approaches in various branches of science, indicating that a nonextensive statistical theory, with a power-law type distribution, provides a solution to several anomalous problems in which Boltzmann-Gibbs-Shannon statistical mechanics fails. Nonextensive statistical mechanics has applications in, e.g., cosmology and gravitation, digital image processing, magnetism, complex systems theory, etc.; the number of publications concerning it is huge. There are more than 8000 publications that study or reference the usually called Tsallis statistics (Website 1, 2022).

Herein, in Part I, Chapter One, a general introduction to the physics of manganites is made, especially for the $Pr_{1-x}Ca_xMnO_3$ series. Thus, the main theories that explain the exotic behavior of these manganese oxides are summarized. We have, for example, the double-exchange model (the original and the current one), the model of polarons, the Anderson localization model, and the phase separation scenario in a percolative process. Likewise, we will find a summary of some models for manganites from first principles as well as a current overview of nanostructured manganites.

In Part II, in Chapter Two, before the formalism of nonextensive statistical mechanics is introduced, we will review a historical evolution of the ideas that led to the foundation of the statistical concepts and, especially, to the construction of statistical mechanics. Likewise, it is introduced some entropies successfully applied in different areas of science. Afterward, we will deal with nonextensive statistical mechanics; we will learn the following topics: how Tsallis entropy is postulated, the link between Renyi entropy and Tsallis entropy, pseudo-additivity for Tsallis entropy, escort entropies, generalized mean values, three definitions of temperature in that statistical theory, etc. Furthermore, in Chapter Three, we will learn about an alternative method to calculate the mean values for a compound quantum system $\mathrm{A} + \mathrm{B}$; this method is applicable to *any* theory using a power law, including obviously nonextensive statistical mechanics. Specifically, only for didactic reasons, a two spin-½ system is studied; the same study can be made for spins 2 and 3/2, Mn^{3+} and Mn^{4+} spins, respectively. Then, two procedures are utilized to calculate the mean values of the two spin-½ system: The usual method of partial matrices raised to the entropic index q, ρ_A^q and ρ_B^q, and an alternative method in agreement with the partial trace quantum operation.

In Part III, applications of the topics discussed are explained. Thus, in Chapter Four, for $Pr_{1-x}Ca_xMnO_3$ with $x = 0.30$–0.50, it is introduced a phenomenological and specific model by using the alternative method along with the mean-field approximation. This model works well in studying specific aspects of manganites. Chapter Five summarizes the matters we will deal with.

Furthermore, it is introduced three appendices. In Appendix A, magnetoresistance and the magnetocaloric effect and their applications will be explained. In Appendix B, we will establish several properties of the q-logarithm function (for $q = 1$, we recover the natural logarithm function). Finally, in Appendix C, we will find out several properties of the q-operations (q-sum, q-difference, q-multiplication, and q-division). Likewise, we will discover how to obtain the skewness and kurtosis for both the standard Gaussian distribution and the q-Gaussian distribution.

PART I

MANGANITES

CHAPTER ONE

INTRODUCTION TO THE PHYSICS OF MANGANITES

Manganites are materials that are currently being intensively investigated by the scientific community; special attention is given to perovskite manganites, which have rich and complex physics at the same time. This is due to the electron-lattice and electron-electron interactions as well as the mixed valence of the manganese ions (Mn^{3+} and Mn^{4+}). These features allow the structural, magnetic and transport properties to be intrinsically related to each other.

Another characteristic, important and unusual, that distinguishes these materials: For certain compositions, in the vicinity of a metal-insulator transition, they present magnetoresistance with very large negative values (it is named colossal magnetoresistance, CMR). Certainly, nanostructures are at the heart of CMR; there are also studies of compounds in which similar nanostructures have not been fully understood, for example, high-temperature superconductors and Eu-based semiconductors. In addition to colossal magnetoresistance, manganites have metallic phases with a conduction band having polarized spin (semi-metallic behavior); this last fact is promising for potential technological applications.

A Brief Historical Outlook

Historically, for the first time, the perovskite manganites $La_{1-x}M_xMnO_3$ (M = Ca, Sr, Ba) were studied in the 50s. Because of their experimental studies, using polycrystalline samples, G. H. Jonker and J. H. Van Santen as well as E. O. Wollan and W. C. Koehler are renowned researchers. However, we also have relevant theoretical investigations from C. Zener, P. W. Anderson, and H. Hasegawa as well as P. G. de Gennes. These last researchers developed the double-exchange model to explain the antiferromagnetic-insulator and metal-ferromagnetic states at certain concentrations x. Yet investigations about manganites only boosted with the discovery of colossal magnetoresistance in the middle of the 70s. If we use the parameter $MR = \frac{R_{(0)}-R_{(H)}}{R_{(H)}}$, we can find magnetoresistance values of 127 000 % at 77 K ($R_{(0)}$ is the zero-field resistance, and $R_{(H)}$ is the resistance having an external field). Besides, if magnetoresistance is expressed via the alternative parameter $mr = \frac{R_{(0)}-R_{(H)}}{R_{(0)}}$, it is about 99 %. In the $Nd_{0.7}Sr_{0.3}MnO_{\delta}$ thin films, we can find values of 10^6 %.

General Formula

The most interesting perovskite manganites have the general formula $R_{1-x}M_xMnO_3$, where R is a rare earth metal, trivalent, and M is an alkaline earth metal, divalent, which can be Ca, Sr, Ba, or Pb. The last formula is usually used when we have electron-hole doping. For instance, $LaMnO_3$, the mother compound of manganites, has x =0 (no doping) and 100 % of Mn^{3+}; the calcium doping, $La_{1-x}Ca_xMnO_3$, provokes the pairing of the Mn^{4+} ions, that is, x is associated with divalent cations.

Similarly, if we make an electron doping, we will use $R_xM_{1-x}MnO_3$; for example, $CaMnO_3$ has $x = 0$ (no doping) and 100 % of Mn^{4+}. The lanthanum doping, $La_xCa_{1-x}MnO_3$, causes the pairing of the Mn^{3+}ions, i.e., x is now associated with trivalent cations. An alternative notation for the general formula, with hole doping, is $R_{1-x}M_xMn^{3+}_{1-x}Mn^{4+}_xO_3$. Besides, electron doping can be performed on $LaMnO_3$, using tetravalent cations like Ce, Sn, and others; so, the construction of diode-type structures p-$La_{2/3}Ca_{1/3}MnO_3$/n-$La_{2/3}Ce_{1/3}MnO_3$, promising in spintronics, would be feasible.

Crystal Structure

The crystal structure of manganites resembles cubic perovskite, which are mixed oxides with stoichiometry ABO_3. The perfect cubic perovskite has a formula per unit cell (5 atoms); one feature of these materials is the wide variety of substitutions accepted by its crystal structure. The site A can be occupied by almost 25 chemical elements and the site B by almost 50 (but not all perovskites are oxides: F, Cl, or Br can also take the place of O in the structure).

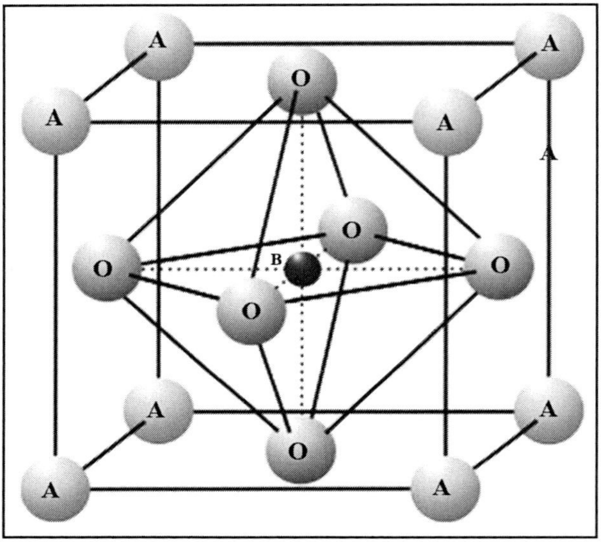

Figure 1.1. Cubic perovskite structure ABO_3. The gray spheres at the corners represent either rare earth or alkaline earth (A); the black sphere in the body center, a manganese atom (B); and each gray sphere in the middle of each cube face, an oxygen atom (O). This figure was modified from Website 2 (2022).

Figure 1.1 shows the perfect cubic-perovskite structure; concerning oxygen (O), a cation A has dodecahedral coordination, whereas the cation B, octahedral coordination. In this structure, the cation B is surrounded by six oxygen anions, forming the BO_6 regular octahedra so that the B–O–B bond angle is 180°. The deviation from the B–O–B bond angle, for Mn, results in an octahedral tilting as well as a minor overlapping between orbitals; this causes the system to lose its metallic characteristics.

Below a given temperature, compounds ABO_3 have structural transitions reducing their symmetry. In general, perovskites ABO_3 have three types of fundamental distortions: **1)** Displacements of the cations from their equilibrium positions. **2)** The BO_6 octahedral distortion due to, e.g., the presence of a Jahn-Teller ion, such as Mn^{3+}; what is provoked by the Jahn-Teller effect—named after Emil Jahn and Edward Teller by the 1937 theorem—which can be static or dynamical. The former Jahn-Teller effect occurs when distortion is large, and permanently affects the molecular geometry; in the latter Jahn-Teller effect, distortions are generally smaller. Two possible distortions are associated with the Jahn-Teller effect: Q_2 is an orthorhombic distortion and Q_3, tetragonal. The change of distances for the Mn–O bond is the main result of such distortions. **3)** Cooperative rotation of the BO_6 octahedra, because of the reduced size of the cation A. The BO_6 octahedral rotation is known as the $GdFeO_3$ rotation; it has the space group Pnma.

In the formation of perovskites, a prerequisite is that the cations A and B have the coordination structures already mentioned. Concerning oxides, this precondition sets lower limits for the radii of these cations [A (0.90 Å) and B (0.51 Å)]. Thus, a parameter that governs the crystallography of manganites can be defined, which was done by Goldschmidt et al. in 1926 in a German journal. They expressed their experimental observations in the famous formula of the tolerance factor t, which has the following expression:

$$t = \frac{1}{\sqrt{2}} \frac{r_A + r_O}{r_B + r_O}, \qquad (1.1)$$

where r_A, r_B and r_O are the ionic radii of the respective elements. Then, $t = 1$ corresponds to a maximum, perfect closely-packed cubic structure. For $t \neq 1$, at low temperatures, manganites have either a rhombohedral symmetry (with space group D_{3d}^6) or orthorhombic symmetry (with space group D_{2h}^6).

Pyrochlore Manganites

Nowadays, the most studied manganites are the ones with a perovskite structure. However, there are also other crystal structures for them. A well-known example is manganites with general formula $X_2Mn_2O_7$ (X = Tl, In, Y). They have a pyrochlore-type crystal structure (with space group $O_h^7, Fd\bar{3}m$). This type of arrangement consists of two sublattices, which respectively have two types of crystallographically different oxygen. Below, the symbol O(1) will indicate the oxygen in the sublattice (1); similarly O(2). Thus, the first sublattice is $MnO(1)_{6/2}$, which forms octahedrons of oxygen, with manganese in the center, the same as in perovskites, connected to each other in the form of rings. The second sublattice is $X_2O(2)$, which passes between the rings formed by the $MnO(1)_{6/2}$ sublattice. These structures have a global cubic symmetry.

The $Tl_2Mn_2O_7$ manganite is important, as it presents colossal magnetoresistance around the FM-PM transition temperature, $T_C \sim 120$ K. Finally, a well-known fact in the scientific literature: Many pyrochlore manganites have a high geometric-magnetic frustration.

Electronic Structure

For an isolated 3d-transition metal, five degenerate orbital states are available for the 3d electrons with $l = 2$ (the energy levels of the Mn^{3+} orbitals are shown in Figure 1.2). The presence of an oxygen octahedron removes this degeneracy. The five d orbitals are shifted into two degenerate groups: triply degenerate t_{2g} orbitals (d_x, d_z, and d_{zx}) and doubly degenerate e_g orbitals ($d_{x^2-y^2}$ and $d_{3z^2-r^2}$). However, the Mn ion is immersed in a non-cubic crystal field, which removes the degeneracy of the t_{2g} and e_g orbitals. The t_{2g} electrons have a localized character and the e_g electrons, an itinerant character; after an explanation of the double-exchange phenomenon, this will be clearer. For the Mn^{3+} and Mn^{4+} ions, the atomic correlations guarantee a parallel alignment of spins (Hund's first rule). The Mn^{3+} ion presents the electron configuration $3d^4, t_{2g}^{3\uparrow}e_g^{\uparrow}$ with $S = 2$ (everywhere in this book, S stands for spin and S, entropy). According to Hund's rule, this cation can accept the fourth electron as being antiparallel to the other three $t_{2g}^{3\uparrow}$ electrons. But because the exchange energy is around 2.5 eV, which is higher than the crystal field splitting $\delta \sim 1.5$ eV, the fourth electron aligns parallel to the $t_{2g}^{3\uparrow}$ electron, in the e_g orbital (in the lanthanum cobaltite $LaCoO_3$, Co^{3+} has six 3d-electrons in the $t_{2g}^{3\uparrow,3\downarrow}$ orbital). The Mn^{4+} ion presents $3d^3t_{2g}^{3\uparrow}$ with $S=3/2$.

Main Theories Seeking to Unravel the Physics of Manganites

In the scientific literature, there are several theories that explain the physical properties of manganites. Then, summarizing the best known, we have the following ones: the original model of double exchange, the current model of double exchange, the model of polarons, and the Anderson localization model. In the next section, we will introduce an important model appearing in the scientific literature: the phase separation scenario. In Figure 1.3, we have a conceptual diagram about these theories.

We begin with an introduction to the original model of double exchange, wherein the magnetic properties are dominated by short-range exchange interactions between the spins of Mn—in 1996, E. L. Nagaev proposed a completely different theory, i.e., the charge carriers are holes in the p-band of oxygen.

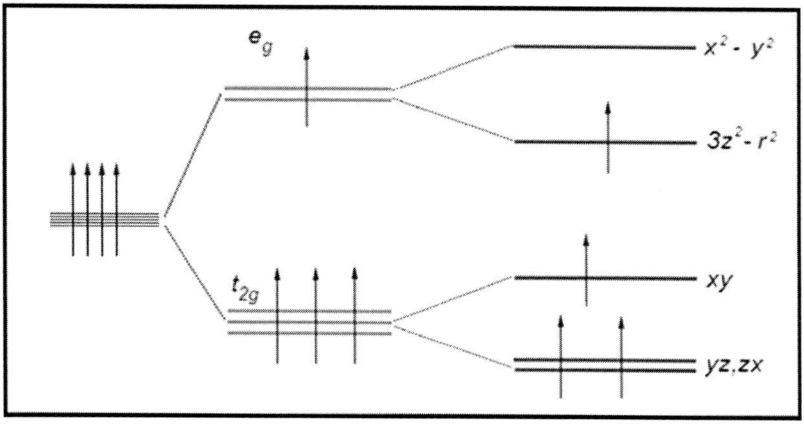

Figure 1.2. Energy Levels of Mn^{3+}. On the left side, we have the case of a free ion; at the center, the levels in the presence of an oxygen octahedron (cubic crystal field); and on the right side, the levels in the presence of a tetragonal crystal field.

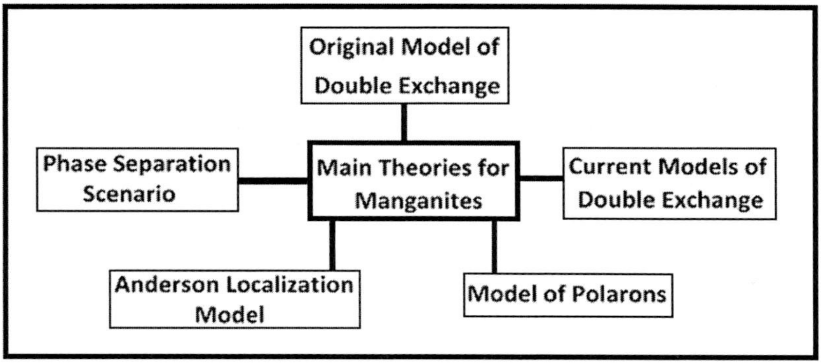

Figure 1.3. Conceptual diagram showing the main theories that explain the properties of manganites.

Such properties are relevant between two spins of manganese separated by an oxygen atom, and they are controlled by the overlapping between manganese's d-orbitals and oxygen's p-orbitals. An interesting case is Mn^{3+}–O–Mn^{4+} in which the Mn ions can exchange their valence, through a jump of the e_g electron of Mn^{3+} into the p-orbital of O, and, simultaneously, another jump of an electron from O to the empty e_g orbital of Mn^{4+}. This mechanism, called double exchange, was proposed for the first time by the American physicist Clarence Melvin Zener in 1951; it ensures a strong ferromagnetic-type interaction.

Later, in 1955, the American and Nobel Prize in Physics Philip Warren Anderson and the Japanese researcher H. Hasegawa enhanced the ideas of C. M. Zener by introducing the probability of the e_g electron to be transferred from Mn^{3+} to its neighbor Mn^{4+}. Thus, we have the electron mobility formula: $T = T_0\mathbf{cos}(\Theta/2)$, where T_0 symbolizes the hopping integral, and Θ represents the angle between spins of localized t_{2g} electrons. $\Theta = 0$ corresponds to a ferromagnetic arrangement, with maximum electron mobility, i.e., the probability of the jumps is maximum in $\Theta = 0$; it is obtained, therefore, a metal ferromagnetic behavior. Likewise, $\Theta = \pi$ corresponds to an antiferromagnetic arrangement; in this case, the conduction e_g-electrons mobility is canceled, that is, we have an antiferromagnetic insulator behavior.

The quantum model analogous to the earlier semiclassical vision was presented in 1972 by the Japanese researchers Kenn Kubo and Nagao Ohata. The competition between the double-exchange ferromagnetism and the superexchange antiferromagnetism gives rise to complex magnetic phase diagrams (superexchange is an indirect exchange interaction in which the separation between ions is greater than, for example, the double exchange). That double-exchange model has been utilized to explain the colossal magnetoresistance phenomenon (however, some authors claim that certain materials—such as pyrochlore manganites and chromium spinels—present colossal magnetoresistance, but not double exchange). For example, the Japanese Nobou Furukawa in 1995, using double exchange, has shown that there is a relationship between resistivity, ρ, and magnetization, M, with a small $M/M_{\mathrm{sat.}}$:

$$\frac{\rho}{\rho_0} = 1 - \mathrm{C}\left(\frac{M}{M_{\mathrm{sat.}}}\right)^2, \qquad (1.2)$$

where ρ_0 is the resistivity without application of a magnetic field; C, a constant; and $M_{\mathrm{sat.}}$, the saturation magnetization.

At present, there are modern models of double exchange. The original meaning of the term *double exchange* has been, in some way, distorted because many scientific studies do not consider the degrees of freedom for oxygen, which exerts a bond between the manganese ions; they consider only manganese Hamiltonians, which can give origin to ferromagnetic phases. Similar models disregarding oxygen are also called double exchange. Usually, in this model, it is indicated that the e_g electron jumps from Mn^{3+} to Mn^{4+} with the extraordinary fact that the electron memorizes what its value of spin is. The e_g-electron transfer, from Mn^{3+} to Mn^{4+} due to the double exchange, is the basic mechanism of electrical conductivity. In those manganites with a strong double exchange, the $e_g$electrons are in a ferromagnetic phase, being $x \sim ⅓$ (concerning $R_{1-x}M_xMnO_3$). Meanwhile, the double-exchange model is known to be incomplete in explaining the complex behavior present in the manganese oxides. However, in 2001, by using computer calculations for the double-exchange model, Atsuo Satou and Masanori Yamanaka succeeded to show that the ground state of manganites spontaneously exhibits a self-similarity, that is, a fractal shape; specifically, they found the Cantor set.

The model of polarons is another much-studied model; there is evidence for the existence of polarons in $La_{1-x}Sr_xMnO_3$ with x = 0.3 and 0.4. It has been observed a strong and reversible change of the electronic structure above the Curie temperature. The localization of the charge and an increase in the Mn magnetic moment, as well as Jahn-Teller distortions, are evidence for the formation of the Jahn-Teller polarons. This phenomenon only appears when the sample is heated above the Curie temperature T_C. Those polarons are consistent with a phase separation scenario. However, according to some investigators, the fact that polaron theoretical models are valid at high temperatures—outside the temperature range in which the colossal magnetoresistance occurs—greatly limits these models (Dagotto 2005). Even the complex and small clusters with CE-phases could not be interpreted as a gas of polarons; a CE-phase is the one with charge ordering and mixture of the C- and E- type antiferromagnetism in the insulating state.

Finally, we have the Anderson localization model, which was utilized to explain the insulating state above T_C; however, the enormous disorder, necessary to obtain the localization in the major densities, is an obstacle to the success of the theory. Moreover, this model does not consider phase competition, a mechanism apparently necessary to explain colossal magnetoresistance. Furthermore, for illustrative purposes, it must be mentioned that this model emerged in 1961 to study a magnetic impurity immersed in a metal matrix. Five years later, John Robert Schrieffer and Peter Adalbert Wolff showed that the Kondo Hamiltonian could be derived from the Anderson model within an appropriate limit. The Kondo model is named after the Japanese theoretical physicist Jun Kondo, who also has studied two-band superconductors and the anomalous Hall effect. Even more, in general, J. Kondo has contributed to understanding the physics of many-body systems and, especially, he has shed light on the study of magnetic phenomena within solid state physics.

Phase Separation Scenario

This scenario has been proposed to explain the complex behavior of manganites. There are dozens of experiments supporting this scenario (by using techniques like electron microscopy, electron transport, nuclear magnetic resonance, neutron diffraction, synchrotron X-ray diffraction, etc.). Thus, there have been observed phases' competition and coexistence. Specifically, it has been reported some mixtures of the following phases: metal and insulator (electronic); ferromagnetic, antiferromagnetic, and paramagnetic (magnetic); and charge and orbital orderings. Phases' competition and coexistence would also arise in superconducting cuprates. There is evidence for the competition between a superconducting phase and a mixture of charge-stripe ordering in $La_{1.6-x}Nd_{0.4}Sr_xCuO_4$.

Even so, some researchers are cautious about the success of this scenario and suggest additional research concerning it; that prudence is due to all basic characteristics of the phase separation are not known currently, for example, the dynamical behavior. Consequently, we have at least three models concerning what the origin of this phase separation is: **1)** electronic-driven phase separation scenario, **2)** phase separation scenario induced by disorder, that is, in a concomitant percolative process, and **3)** strain-driven phase separation scenario.

In respect of the first model, there is theoretical research showing the existence of phase separation between regions rich and poor in the content of charge carriers, that is, with different electron densities. For example, in this context, the one-orbital model and the two-orbital model can be used to study phase separation. The calculation of the e_g-electron density, as a function of the chemical potential, results in discontinuities in the electron density for some values of the chemical potential. This fact is interpreted as the existence of inhomogeneous ground states, separated into regions with different charge densities. Meanwhile, there is also another phase separation scenario in which the electron densities are equal.

With respect to the second model, in a first-order transition in the absence of disorder, the system gets confused and does not know if it is insulating or metallic. However, when the disorder is dominant, fine clusters of the two phases are formed. The competition of the two phases, in the presence of disorder, will finish in larger clusters with equal density. In this second model, the metal-insulator transition and the associated colossal magnetoresistance occur in a percolative and concomitant process of the metal ferromagnetic domains. The aforementioned first-order transition is caused by different charge and magnetic orderings of the phases in competition. Furthermore, the coexistence of metal clusters and charge-ordered clusters has been experimentally verified; the value of the applied magnetic field determines the size and structure of the previously mentioned clusters.

Concerning the third model, strain is assumed to be the driving force of the phase separation. In this model, lattice distortions are important for manganese oxides. The elementary assumptions in the model are **(a)** the system is metallic when strain is absent, and the same system is insulating when strain is present; **(b)** the system has Jahn-Teller lattice distortions modifying the elastic energy of the system.

Consequently, we can affirm that, nowadays, phase separation is an important process in the physics of manganites. Inherent defects in the sample manufacturing process, such as variations in the oxygen content as well as twinning, could be also behind this phenomenon of phase separation. However, though the phase separation scenario can be separated into disorder-driven or intrinsic electronic, the truth is probably a mixture of both. Therefore, whatever the cause that triggers the formation of clusters and percolation, phases' competition and coexistence constitute the fundamental theory behind (I thank Prof. Elbio Dagotto for his comments that helped me improve this paragraph).

On the other hand, $La_{5/8-y}Pr_yCa_{3/8}MnO_3$ is considered a prototype material to study the phase separation; for y = 0, we have $La_{5/8}Ca_{3/8}MnO_3$, which has a ferromagnetic metal state at low temperatures. For y = ⅝, we have $Pr_{5/8}Ca_{3/8}MnO_3$, which has an insulating state with charge ordering. Likewise, typical characteristics of the spin-glass state have been found in manganites with phase separation (however, in the $Eu_{0.5}Ba_{0.5}MnO_3$ single crystals, we have a spin-glass state with orbital ordering but without phase separation). In the $(La_{0.25}Nd_{0.75})_{0.7}Ca_{0.3}MnO_3$ ceramic, the origin of magnetic frustration and the spin-glass behavior in the phase separation are very complex; the coupling of a set of interacting magnetic clusters would be sufficient to explain the spin-glass behavior. This behavior can also be attributed to the interaction of clusters within the separate phase state.

Besides, it is also convenient to report upon the model that considers colossal magnetoresistance as a Griffiths singularity. Then, the magnetic transition provokes colossal magnetoresistance; this transition can be considered as some kind of percolation, but these ideas have to be seen within the context of the Griffiths phase, which is a peculiar phenomenon of the disordered systems. Specifically, the Griffiths singularities can be used to explain anomalous aspects of susceptibility and specific heat in the $La_{0.7}Ca_{0.3}MnO_3$ single crystal. In classical statistical mechanics, the

Griffiths singularities are considered essential singularities and, therefore, they have no effect on the static properties of the system.

Furthermore, there are two types of colossal magnetoresistance at different temperature intervals. In addition to the standard colossal magnetoresistance, next to T_C, there is another colossal magnetoresistance at low temperatures. Besides, manganites $(Nd_{1-y}Sm_y)_{0.5}Sr_{0.5}MnO_3$ and $Pr_{0.70}Ca_{0.30}MnO_3$ have been experimentally used to provide evidence in this regard. Likewise, the neutron diffraction technique, with inelastic scattering, can be used to show the discontinuous nature of the metal-insulator transition induced by an external field. Thereupon, Monte Carlo simulations can be also used to study half-doped manganites.

Charge-Ordering Phase Transition

This one has been known since 1939, when the Dutch chemist Evert Johannes Willem Verwey described, in ferrous ferrite (Fe_3O_4 with spinel structure), the transition from a state with charge ordering to a disordered state, upon exceeding a particular temperature. Verwey assumed that the Fe^{2+} and Fe^{3+} cations are alternately arranged. In the scientific literature, such a transition is known as the Verwey transition. In some transition-metal oxides, this system has also been reported; however, it is in manganites that it has attracted the attention of researchers.

The charge ordering in manganites was observed for the first time in 1955, by E. O. Wollan and W. C. Koehler, who used neutron diffraction in $La_{1-x}Ca_xMnO_3$ with $0 < x < 1$. In 1985, the charge ordering has been also studied by Jirak et al. by using neutron diffraction in $Pr_{1-x}Ca_xMnO_3$ with $0 < x < 1$. This phase transition in manganites could be due to the Coulombic interaction, which prevails over the kinetic energy; nonetheless, there is also evidence for the influence of the electron-phonon interaction. The insulating state presents antiferromagnetic stripes of static charges; in $La_{1-x}Ca_xMnO_3$ (with x = 1/2, 2/3, 3/4, and 4/5), it has been discovered pairs of octahedral stripes $Mn^{3+}O_6$ separated by octahedral stripes $Mn^{4+}O_6$.

Next, as a didactic example, we can mention $La_{0.5}Ca_{0.5}MnO_3$, for which it is possible to obtain the Mn^{3+} and Mn^{4+} charge ordering below the critical temperature $T_{CO} = 150$ K (therefore, it has a crystal structure with two classes of the MnO_6 octahedra). In other words, there is a phase transition below T_{CO}, known as Wigner crystallization or Verwey transition. Below T_{CO}, the extra electron of Mn^{3+} orders alternately onto the Mn^{4+} ions (forming a rock salt-type arrangement); besides, above T_{CO}, the sizes of the MnO_6 octahedra are equal, with a manganese valence state on average $Mn^{3.5+}$. The larger octahedra have Mn^{3+} at the center; the smaller, Mn^{4+}. Then, $Nd_{0.5}Sr_{0.5}MnO_3$ and $Pr_{0.6}Ca_{0.4}MnO_3$ can be also considered representative materials in studying the charge ordering in manganites. This ordering provokes them an extraordinary group of properties such as sensitivity to the following: average ionic radius of cations A ($\langle r_A \rangle$), pressure, electric and magnetic fields, and chemical and isotopic substitutions. Furthermore, concerning average ionic radius $\langle r_A \rangle$, it is possible to create another variable: the degree of disorder of the site A, $\sigma^2 = \langle r_A^2 \rangle - \langle r_A \rangle^2$.

Orbital Ordering

This one can be found in localized 3d systems in which the cooperative Jahn-Teller distortions are important because of the strong hybridization between the 3d and oxygen-2p electrons. This ordering has been found in titanates, vanadates, and magnetite.

Specifically, some manganites exhibit an e_g orbital ordering, which is attributed to a very strong link between electrons with the Jahn-Teller phonon. We must consider that this ordering can promote or disfavor double-exchange and superexchange interactions according to the direction of the orbital. The work of van den Brink, Khaliullin, and Khomskii (2004) is an excellent review dedicated exclusively to the orbital ordering in manganites. Additionally, the review article of Khomskii (2001) is interesting.

The $Pr_{1-x}Ca_xMnO_3$ Series

This compound for x = 0.3–0.5 presents a transition from a paramagnetic insulator phase (PI) to an insulator charge-ordering phase (ICO), with $T_{CO} = 220$ K. The ratio 1:1 of Mn^{3+} and Mn^{4+} occurs at $T_{CO} \sim 230$ K. The change toward the charge ordering state is accompanied by an alteration in the lattice parameters (~ 0.6–2.4 %); below T_{CO}, it has been found the phases antiferromagnetic insulator (AFI) and canted-antiferromagnetic insulator (CAFI). For $0.2 < x < 0.3$, at low temperatures, the exotic ferromagnetic insulator phase (FI) can be found. For x = 0.3, in the presence of an external magnetic field and at low temperatures, it occurs a sudden transition from a CAFI phase, with high resistivity, toward a ferromagnetic metal phase (FM) showing a strong change in resistivity. This type of change can reach ratios as high as 10^{11} in accordance with the applied doping. According to A. M. Haghiri-Gosnet and J. P. Renard (2003), this metal-insulator transition is considered an abrupt jump of the resistivity without a singularity in the magnetization, dependent on the temperature, which would be proof of the percolative nature of the transition. Together with this first-order transition, strong changes occur in the lattice parameters as well as noticeable hysteresis effects do.

A remarkable feature of these compounds is the irreversibility that presents the FM phase, that is, it is observed even after removing the external magnetic field. The AFI phase cannot be recovered; as an example, we have $Pr_{0.7}Ca_{0.3}MnO_3$. The (Pr,Ca) series has a low bandwidth and it is interesting because the metastable insulating state can become a metallic state due to various external agents, in addition to the magnetic field: electrical fields, high pressures, exposure to visible light or X-rays. There is an important difference between the magnetization curves, depending on whether the sample was measured by using zero field cooling (ZFC) or by using field cooling (FC); this difference depends on several factors, including the value of the applied magnetic field. For example, in the $Pr_{0.7}Ca_{0.3}MnO_3$ manganites, the two magnetization curves (ZFC and FC) are almost identical with magnetic fields applied above 0.5 tesla.

A Summary of Some Models for Manganites from First Principles

Unlike the models already introduced, these ones start from a first-principles Hamiltonian, which contains all relevant interaction terms:

$$H = K_{\text{kin}} + H_{\text{Hund}} + H_{\text{AFM}} + H_{\text{el-phon}} + H_{\text{el-el}}\,, \qquad (1.3)$$

where K_{kin} is the kinetic term; H_{Hund}, the Hund interaction; H_{AFM}, the antiferromagnetic contribution; $H_{\text{el-phon}}$, the electron-phonon interaction; and $H_{\text{el-el}}$, the electron-electron interaction. Details of this Hamiltonian appear explicitly in the review authored by Dagotto, Hotta, and Moreo (2001). Specific models for manganese oxides arise when some approximations are considered:

a) **Electron gas model**. It retains only the kinetic term; it is a useful model in the ferromagnetic region, wherein the static Jahn-Teller distortion does not occur. The t_{2g} electrons have a localized character, but the e_g electrons can move through the 2p-orbital of oxygen.

b) **One-orbital model**. This one illustrates the colossal magnetoresistance; it abandons the $H_{\text{el-phon}}$ and $H_{\text{el-el}}$ terms. The additional disregard of the orbital degrees of freedom leads to a ferromagnetic Kondo model (or double-exchange model for one orbital). Since the e_g electrons do not exist in $CaMnO_3$, the G-type antiferromagnetic ordering arises. To explain this, the Heisenberg term between the localized t_{2g}-electrons must be considered. Evidently, the (e_g electrons) orbital ordering of manganites cannot be explained by this model.

c) **Model $J_H = \infty$**. This limit of the Hund coupling J_H is possible without loss of physical interpretation. In this context, the e_g-electron spin aligns perfectly with the t_{2g}-electron spin, reducing thus the number of degrees of freedom.

d) **Phononic Jahn-Teller model**. This approach does not consider the $H_{\text{el-el}}$ term, but it maintains H_{Hund} finite; consequently, a model with (active) spin degrees of freedom is obtained. The Hamiltonian is given by

$$H_{JT} = K_{\text{kin}} + H_{\text{Hund}} + H_{\text{AFM}} + H_{\text{el-phon}} \, . \quad (1.4)$$

To find a solution to H_{JT}, numerical methods, such as Monte Carlo and relaxation methods, have been used.

Due to the complexity and difficulty of considering all the terms, approximate models of the first-principles Hamiltonian naturally emerge.

The complications are too many, so it is usual to investigate manganites only from approximate models.

Nanostructured Manganites

Since the last decade, research on nanomagnetism has boomed because the nanostructured materials have applications in digital electronics. Specifically, these submicroscopic materials are part of permanent magnets, telephone GPS, and computer devices. Nowadays, carbon nanotubes are the best-known nanostructured materials. However, there are also other types of nanostructures such as titanium dioxide, graphene, gold, and quantum dots.

Concerning manganites, we can find theoretical and experimental research. Thereupon, we have as follows: manganite nanocrystals synthesis by the hydrothermal method; studies of a magnetic dead layer in manganite nanoparticles; studies of their structural and magnetic properties by utilizing techniques like powder X-ray diffraction, transmission electron microscope, and field-emission scanning electron microscope accompanied by magnetic measurements.

Consequently, for example, the $La_{0.7}Ca_{0.3-x}Sr_xMnO_3$ manganites can be manufactured via four methods: the microwave-assisted sol-gel Pechini method, the Pechini sol-gel synthesis, the solid state reaction with a planetary ball mill, and the solid state reaction with a stirred ball mill. Also, there are other materials such as the $La_{0.425}Pr_{0.2}Ca_{0.375}MnO_3$ nanodisk arrays, which are fabricated from epitaxially grown thin films. The aforementioned material could be used for future magnetic refrigerators. Besides, by using a combustion method, nanocrystalline $La_{0.9}A_{0.1}MnO_3$ powders can be synthesized (A = Li, Na, K). Likewise, the $Pr_{0.5}Sr_{0.5-x}Na_xMnO_3$ manganites present interesting magnetic and magnetocaloric properties. Furthermore, nanocomposites of lanthanum strontium manganite are interesting for solid oxide cells. Finally, silver-doped lanthanum manganite can be used in selective chemotherapy, thus avoiding the danger of radiotoxicity.

Potential Technological Applications

Magnetic multilayers with manganites have possible implementations in the spintronics industry, due to their ability to inject polarized spin current; the bilayer Fe_3O_4 and $La_{0.7}Ca_{0.3}MnO_3$ is interesting because of the fascinating electrical and magnetic properties of the spinel and perovskite structures, respectively. Besides, the magnetocaloric effect in manganites can be used in the construction of magnetic refrigerators. Concerning the growth of manganite samples, A. M. Haghiri-Gosnet and J. P. Renard (2003) consider that it is mastered from all viewpoints, e.g., crystallinity, effects of strain on the magnetic anisotropy, magnetoresistance, etc. These last authors also consider that two objectives that scientists should prioritize are the following:

(a) Build tools for industry, such as elements of sensors using magnetoresistive random-access memory (MRAM), position sensors, contact-less potentiometers, bolometers (these last are non-coherent thermal detectors in which an incident light causes changes in the temperature of an absorber material). Concerning MRAM sensors, dry etching is a very important technical procedure for their nanofabrication, allowing the transfer of a nanopattern with high anisotropy, selectivity, and uniformity. So, the nanopattern is less damaged—in semiconductor technology, the dry etching is a process that uses chemical corrosion, through an ionized gas, to remove unprotected portions in the layer of a given material.

(b) Extending the knowledge of the physics of these materials in geometries such as artificial heterostructures, for which new functionalities can be achieved. Additionally, concerning this last item, research has been extended to other oxides that present the phenomenon of colossal magnetoresistance like the $La_{2-2x}Sr_{1+2x}Mn_2O_7$ multilayer manganites—which consists of Mn bilayers separated by nonmagnetic layers—and Sr_2MoFeO_6, too. This last double perovskite is unstable in the presence of moisture, and it needs a high deposition temperature, $T \sim 900$ °C. This perovskite would be a candidate material for tunneling devices, that is, the electron tunneling between ferromagnetic electrodes. The geometrically frustrated pyrochlores, like $Nd_2Mo_2O_7$, can be also mentioned; nonetheless, optimism regarding pyrochlores and double perovskites has decreased.

Furthermore, the $La_{1-x}Sr_xMnO_3$ series is important when the growth process is easy to control, due to its stability, and because the material can be easily nanopatterned. Spin filters using $La_{1-x}Sr_xMnO_3$ thin films could increase their spin polarization. Furthermore, there are proposals for the construction of a magnetic handle for antiferromagnetic manganites.

This chapter showed that manganites possess several characteristics, which are interesting for basic and applied physics. There is no single theoretical model explaining the very complex behavior of these oxides. The various degrees of freedom (spin, orbital, charge, and lattice) interact in a synergic way, leading to a complexity intrinsic to these manganese oxides. Thus, nonextensive statistical mechanics is a guide to studying these materials. In Chapter Four, we will deal with the $Pr_{1-x}Ca_xMnO_3$ series with x = 0.35–0.50; we also will show a phenomenological model, from which emerges nonextensive quasi-probability distributions. But before, in Chapter Two, we will introduce the theoretical framework of nonextensive statistical mechanics and, in Chapter Three, we will learn a new method to calculate the mean values of compound systems.

PART II

NONEXTENSIVE STATISTICAL MECHANICS

CHAPTER TWO

NONEXTENSIVE STATISTICAL MECHANICS

This statistical theory is part of a plethora of propositions searching for generalizing the standard statistical mechanics (also known as Boltzmann-Gibbs-Shannon statistics). For instance, there are theories utilizing, respectively, the Sharma-Mittal, Renyi and Kaniadakis entropies. As a result of its application to multiple and varied systems, nonextensive statistical mechanics has been widely refined since its proposal in 1988. At the moment, we can find over 8000 scientific publications studying or referencing this theory (Website 1, 2022). This statistical mechanics, based on a power-law type distribution, helps us understand some of the anomalous phenomena in statistical physics.

Before introducing the formalism of nonextensive statistical mechanics, we will show a historical evolution of some ideas that led to the construction of important concepts in statistics and, especially, in statistical mechanics (however, we also will include the main out-of-statistics contributions of the respective science creators. See the historical review by D. Flamm 1997). Afterward, we will show several entropies successful in the respective areas in which they are applied. Then, firstly, we have, chronologically, the following events.

Historical Events

In 1738, the Swiss mathematician and physicist Daniel Bernoulli introduced a derivation of the gas laws from a model of pool balls. He considered a gas was formed by a large number of small particles moving quickly. Besides, Daniel Bernoulli is known to have formulated the mathematics necessary to understand the behavior of the water wheels and water pumps, too.

In 1763, the work of the English statistician and Presbyterian minister Thomas Bayes was published; it is a special case of Bayes' theorem. This one initiates the Bayesian probability theory usually known as *a posteriori* probability theory. Specifically, that theorem allows us to calculate the probability of an event occurring when we know beforehand data on that event.

In 1814, the French mathematician Pierre-Simon Laplace published *Essai philosophique sur les probabilities* (*A Philosophical Essay on Probabilities*), an important book that made known the calculus of probabilities. The *a priori* probability is known as the Laplace probability distribution. Besides, this French polymath is known for inventing the Laplace transform and the Laplace equation. Likewise, in astronomy, between 1798 and 1827, he would publish five historical books named *Traité de Mécanique* (*A Treatise of Celestial Mechanics*). The two first volumes dealt with methods to calculate the movement of the planets and even to study the tides. Furthermore, it is famous the anecdote when Laplace is questioned by the former Emperor of the French Napoléon Bonaparte on the role of God in his aforementioned books: He answered it was a hypothesis that he did not need in his theory.

In 1860, the Scottish mathematician and physicist James Clerk Maxwell published the important article *Illustration of the Dynamical Theory of Gases*, in *The Philosophical Magazine*. The innovation introduced by Maxwell was to consider that the molecules of a gas have a random movement. Before this work, it was usual to consider uniform velocities for the molecules of a gas. In other respects, J. C. Maxwell is the creator of the electromagnetic theory since he compiled and explained several physical laws existing in his age, which nowadays are known as Maxwell's four laws. In other words, he laid the foundations for classical electromagnetism.

In 1865, the German physicist and mathematician Rudolf Julius Emmanuel Clausius made known the thermodynamic entropy. R. J. E. Clausius is one of the founders of thermodynamics. In 1850 he had already published his book *On the Moving Force of Heat*, where he had placed the basis of the second law of thermodynamics. Historians of science also credit R. J. E. Clausius with introducing the Virial theorem in 1870 within the article entitled, in English language, *On a Mechanical Theorem Applicable to Heat*. This publication, for a discrete system of particles, relates total kinetic energy to total potential energy, both averaged over time.

In 1872, the Austrian physicist Ludwig Eduard Boltzmann, within the framework of the kinetic theory of gases, introduced and proved the H-theorem. This one explains how a gas, satisfying the Boltzmann equation, evolves toward the thermodynamic equilibrium. In other respects, in 1871, L. E. Boltzmann had already introduced the ergodic hypothesis in the context of his kinetic theory of gases. This one is an essential hypothesis of statistical mechanics, and it is valid for a large number of particles. This hypothesis establishes that in equilibrium the mean of a statistically calculated magnitude is equal to the mean of a large number of measurements taken over time. On the other hand, about his personal life, we know the tragic way in which L. E. Boltzmann decided to pass away, namely, committing suicide.

In 1876, the American scientist Josiah Willard Gibbs published the important article entitled *On the Equilibrium of Heterogeneous Substances*, which led him to be considered one of the founders of physical chemistry. He is also known as the father of formal statistical mechanics and, a pioneer of the maximum entropy principle. Likewise, J. W. Gibbs was the creator of the concepts of statistical ensemble and chemical potential. The former is an ideal situation, which considers a set of virtual copies of a system, which are in the accessible states of the system; the latter is the rate of change in the internal energy concerning the change in the number of moles. In his aforementioned article, J. W. Gibbs also introduced the famous Gibbs paradox, which implies the discontinuous nature of configuration entropy. This one also called conformational entropy deals with the geometric configuration of the single parts of a system.

In 1877, the aforementioned L. E. Boltzmann introduced a probabilistic expression for entropy. Besides, he is known to have contributed to the development of the kinetic theory of gases. Also, by applying probabilistic methods in mechanics, L. E. Boltzmann built a theoretical basis for the phenomenological laws of thermodynamics. He also proclaimed the existence of atoms when this idea was not accepted by the scientific community.

Between 1901 and 1903, independently of L. Boltzmann and J. W. Gibbs, the German-American theoretical physicist Albert Einstein formulated a new classical statistical mechanics, obtaining the same results as they did. His ideas of obtaining statistical mechanics from dynamics, and some disagreements with Boltzmann statistics, have inspired scientists for decades until today in the search for generalized statistical mechanics in which Boltzmann-Gibbs statistics is a particular case. Besides, it must be mentioned that A. Einstein is considered one of the greatest theoretical physicists of all time since he published the special theory of relativity in 1905 and later, the general theory of relativity in 1915. These last theories improved the ones of Sir Isaac Newton, who in 1687 had already published the work that changed the world in the XVII century—*Philosophiae Naturalis Principia Mathematica* (*Mathematical Principles of Natural Philosophy*). Sir I. Newton was the English genius who led and closed the historical stage known as the scientific revolution.

In 1902, the aforementioned J. W. Gibbs reviewed the relationship between the laws of thermodynamics and the statistical theory of molecular motion.

In 1909, the German-Greek mathematician Constantin Caratheodory published *Investigations on the Foundations of Thermodynamics*. In this work, he axiomatically formulated the second law of thermodynamics. His work is a breaking point because, afterward, thermodynamics was not based on empiricism anymore. Besides, C. Caratheodory is known to have contributed to the calculus of variations, geometrical optics, and analysis of real and complex functions.

In 1920, the German physicist Wilhem Lenz proposed the now called Ising model. In 1925, the German physicist Ernst Ising discussed the details of this model for a linear chain. In 1942, the Norwegian-American chemist Lars Onsanger presented the exact solution for the two-dimensional Ising model (in 1968, he was awarded the Nobel Prize in Chemistry, but the prize was due to another contribution to science).

In 1925, the British statistician Ronald Aylmer Fisher—a pioneer in the application of statistics to design scientific experiments—created an alternative information measure, within the theory of statistical inference, to measure uncertainty on an ordered space. Several years later, in 1935, he would introduce his book *The Design of Experiments* in which appears the known randomized experiment *Lady Tasting Tea*.

In 1937, the French mathematician Paul Pierre Lévy introduced the Lévy distributions, which are common in nature. Some experts consider that the classic probabilists have known, for more than 60 years, the distribution of nonextensive statistical mechanics, in the form of infinitely divisible distributions, that is, Lévy distributions. Paul P. Lévy is considered one of the founders of the modern theory of probability.

In 1946, the Soviet mathematician and physicist Nikolai Nikolaevich Bogolyubov introduced *Problems of Dynamical Theory in Statistical Physics*, where he generalizes the Boltzmann equation by using the Liouville equation invariant under time reversal. This work helped to clarify the internal structure of statistical mechanics. Besides, his contributions to mathematical and theoretical physics matter a great deal. Thus, it is recommended to read the 2006 book *Nikolai Nikolaevich Bogolyubov — Mathematician by the Grace of God* by Vasilii Sergeevich Vladimirov.

In 1948, the American mathematician Claude Elwood Shannon created information entropy, which is considered the extent of the information contained in a message. In other respects, he is famous for having placed, when he was 21 years old in 1937, the basis of the digital circuit design theory. Likewise, C. E. Shannon, working as a cryptanalyst, participated during the Second World War, collaborating with the Allies.

In 1951, the American mathematician Solomon Kullback and the American mathematician Richard A. Leibler made known Kullback-Leibler entropy, which is also known as relative entropy or Kullback-Leibler divergence. Kullback's book *Information Theory and Statistics* deals with this divergence. In addition, S. Kullback, as a cryptanalyst, worked in the National Security Agency (NSA); he was involved in inventions such as drum memory and magnetic tape.

In 1957, the American physicist Edwin Thompson Jaynes published his first article on information theory. Their importance comes from the fact that Jaynes reformulated epistemologically statistical mechanics in terms of probability distributions, which are derived from the use of the maximum entropy principle This simplified the mathematics used and allowed for fundamental extensions of the theory. The aforementioned principle states that, for a system at thermodynamic equilibrium, the suitable probability distribution is the one maximizing entropy. To find that distribution the method of Lagrange multipliers is usually used. This method optimizes a function in the presence of one or more constraints.

Also in 1957, the Soviet mathematician Aleksandr Yakovlevich Khinchin proposed 4 axioms that all entropy must satisfy. Besides, A. Y. Khinchin is known to have contributed to building the modern probability theory; he had already discovered the law of the iterated logarithm in 1924. Likewise, Khinchin's constant is named after him ($K_0 \approx 2.6854520$). This constant is the limit of the geometric mean of the first partial quotients for the continued fraction of almost any real number.

In 1958, the Soviet mathematician Andrei Nikolaevich Kolmogorov formulated the original work, and in 1959 the Russian mathematician Yakov G. Sinai improved what is currently known as Kolmogorov-Sinai entropy. This one is also known as the metric entropy of a dynamical system. Also, A. N. Kolmogorov prepared the ground for the axioms of probability. His book *Foundations of the Theory of Probability* is well known in the scholarly world.

In 1960, the American physicist Herbert Bernard Callen published his famous book *Thermodynamics and an Introduction to Thermostatistics* in which he constructed thermodynamics based on postulates. In other respects, he was involved in research on the construction of the atomic bomb in World War II, namely, the Manhattan project. Besides, in magnetism, H. B. Callen was one of the first scholars to use the thermodynamic Green functions.

In 1961, the Hungarian mathematician Alfred Renyi presented his entropy, nowadays widely used by the mathematical community. His lecture *On measures of entropy and information* is very important and extensively cited in the scientific literature. Besides, his book *Probability Theory* is broadly known in the mathematical academic world. To him is attributed the phrase "a mathematician is a machine for turning coffee into theorems." However, some authors attribute that phrase to his fellow countryman, also a mathematician, Paul Erdös, who published more than 1500 scientific papers in his lifetime.

Also in 1961, the German American physicist Rolf Landauer formulated the principle of Landauer, which deals with the inferior theoretical limit of energy loss of a system. Specifically, this subject is part of the thermodynamics of information processing. In addition, several of his scientific investigations were used on computers. It is recommended to read his writing *Information is physical, but slippery*, in a book edited by Michael Brooks in 1999, *Quantum Computing and Communications*.

In 1967, the American physicist Leo Philip Kadanoff introduced the basic ideas of the theory of scaling. This theory provided the conceptual basis for the renormalization group, a very useful tool for describing critical phenomena. The renormalization group allows us to analyze a transformation of scale, which is obtained by eliminating the degrees of freedom from the system. In other respects, many scholars consider that L. P. Kadanoff had to share the 1982 Nobel Prize in Physics, which was awarded to Kenneth Geddes Wilson only.

In 1974, the American theoretical physicist K. G. Wilson, with his solution to the Kondo problem, consolidated the ideas of group de renormalization. In 1982, K. G. Wilson would win the Nobel Prize in Physics because he had contributed theoretically to understanding critical phenomena and phase transitions. Besides, by using the Monte Carlo renormalization group, he also was interested in the 3D Ising model.

In 1988, the Greek-Brazilian physicist Constantino Tsallis proposed nonextensive statistical mechanics, as a possible generalization of Boltzmann-Gibbs statistics. But, over the years, this version has been modified, appearing the second and third versions. Professor C. Tsallis has been cited more than 10000 times in the scientific literature.

In 1992, the American physicist Steven R. White and the German professor Reinhard M. Noack developed the algorithm known as density matrix renormalization group, which allowed the calculation of the physical properties of quantum systems with a dimensionally large Hilbert space. It is considered the most powerful numerical method for approaching many-body systems.

In 2003, the professor of applied mathematics Christian Beck and the Dutch American physicist Ezechiel Godert David Cohen introduced superstatistics. In this theory, it is shown that Tsallis statistics is recovered from the χ^2 distribution (chi-square distribution). In other respects, the work of E. G. D. Cohen has allowed the development of the helium dilution refrigerator, which attains temperatures lower than 1 kelvin.

Entropies Successfully Applied in Different Areas of Science

Currently, the term 'entropy' has different meanings and applications in various scientific disciplines such as thermodynamics, statistical mechanics, information theory, robotics, pure mathematics (fractals), economics, etc. Below, we have a summary of some important entropies; however, Havrda-Charvat-Daroczy-Tsallis entropy (S_q) will be presented later and its respective formula will be shown in Eq. (2.1).

Thermodynamic Entropy

The term 'entropy' was introduced in 1865, for the first time, by Rudolf Julius Emanuel Clausius, who defined the variation in entropy of a thermodynamic system as $\Delta S = \Delta QT$ (during a reversible process in which a quantity of heat ΔQ is applied at absolute temperature). R. J. E. Clausius gave it the name of entropy due to its Greek origin 'transformation'. However, Clausius' contribution was empirical. The Inaccessibility Theorem of Constantin Caratheodory is precisely what justifies the existence of entropy in thermodynamics.

Boltzmann Entropy

In 1877, Ludwig Eduard Boltzmann, the founder of statistical mechanics, published the paper entitled *On the relation between the second law of the mechanical theory of heat and the probability calculus with respect to the theorems on thermal equilibrium*. In this article, he presented a probabilistic expression for entropy. Currently, Boltzmann entropy is written as $S = k_{\mathrm{B}}\mathrm{Ln}(\Omega)$, where k_{B} is the Boltzmann constant, and Ω is the number of microstates that are consistent with the macroscopic state.

A macrostate is determined by a small number of macroscopic variables, e.g., volume, pressure, and temperature. A microstate, on the other hand, is specified by the coordinates of position and linear momentum of all molecules in the system. L. E. Boltzmann, to obtain a finite number of microstates, introduced cells in the phase space. According to the statistical definition of entropy, the second law of thermodynamics states that disorder in an isolated system tends to grow.

The contributions of Josiah Willard Gibbs—clarifying the concept of entropy, on the basis of phase diagrams in which entropy and temperature are considered as coordinates—caused the following entropy to be deservedly called Boltzmann-Gibbs entropy: S=-$k_B \sum_{i=1}^{N} p_i \operatorname{Ln}(p_i)$, where p_i takes discrete values. p_i is known as probability distribution or probability mass function; i, the index indicating the i-th state; N, the total number of microstates accessible to the system. The Hungarian-American mathematician and physicist John von Neumann developed the quantum formalism for this entropy. For this reason, it is also known as von Neumann entropy.

Shannon Entropy

In 1948, Claude Elwood Shannon, known as the father of information theory, published in *the Bell System Technical Journal* the article entitled *A Mathematical Theory of Communication.* C. E. Shannon defined a measure of information, called entropy: $H = -\sum_{i=1}^{N} p_i \operatorname{Log}_2(p_i)$, which determines the degree of chaoticity of the probability distribution p_i. This parameter is utilized to determine the necessary capacity of a channel to transmit information given. In other words, it measures the average number of bits necessary to represent a set of values, with different probabilities. In the scientific literature, this entropy, together with that of the previous subsection, is sometimes named Boltzmann-Gibbs-Shannon entropy. This is due to the connection between thermodynamic entropy and information entropy.

Quantum Group Entropy

In 1997, the Japanese Dr. of Science Sumiyoshi Abe presented quantum group entropy. This author used the q-calculation (q is the entropic index). In this context, let us define the q-differential as follows: $\frac{df(\alpha)}{d(\alpha;q)} \equiv [f(q\alpha) - f(q^{-1}\alpha)]/[q\alpha - q^{-1}\alpha]$, where $f(\alpha)$ is a function; α, a variable. The previous formula is essential to obtain the following entropic formula: $S = -\sum_{i=1}^{N} \frac{[p_i^{q} - p_i^{-q}]}{[q-q^{-1}]}$, which has invariance $q \leftrightarrow q-1$. The name of quantum group is because, in the study of quantum groups, the formula $[\mathrm{A}]_q \equiv (q^{\mathrm{A}} - q^{-\mathrm{A}})/(q - q^{-1})$ is commonly used; $[\mathrm{A}]_q$ is a q–deformed physical quantity, and A is a physical quantity for $q = 1$.

On the other hand, there is a connection between Havrda-Charvat-Daróczy-Tsallis entropy (S_q) and quantum group entropy (S): $S \equiv [(q-1)S_q - (r-1)S_r]/[q-r]$, where $r = q^{-1}$.

κ-Entropy

In 2001, Giorgio Kaniadakis proposed κ-entropy, which, in discrete variables, can be expressed thus: $S_\kappa(p) = -k_{\mathrm{B}} \sum_{i=1}^{\mathrm{N}} p_i \mathrm{Ln}_\kappa p_i$, where p_i is a probability distribution. The κ-logarithm is defined in this manner: $\mathrm{Ln}_k(x) = (x^\kappa - x^{-\kappa})/(2\kappa)$. Boltzmann-Gibbs entropy is recovered when κ approaches zero. κ-entropy is related to Havrda-Charvat-Daroczy-Tsallis entropy by the following entropic formula: $S_\kappa = (2+2\kappa)^{-1}\alpha^\kappa S_{1+\kappa}^q + (2-2\kappa)^{-1}\alpha^{-\kappa} S_{1-\kappa}^q + \mathrm{c}$, where c is a constant.

Summing up, we have shown the main entropies among a plethora (approximately 20) that currently exist. We conclude there is no single generalized entropy to be used in all areas of science (perhaps there is no universality of entropy). Each entropy is restricted to certain branches of science.

Superstatistics

The pioneers of this theory are the Polish researchers Grzegorz Wilk and Zbigniew Wlodarczyk, who, researching the meaning of the entropic index q in the year 2000, found that the fluctuation of the parameters in the Boltzmann exponential has to do with the mentioned index (e.g. fluctuation of temperature, for $q > 1$). The superstatistical theory was formally proposed in the paper entitled *Superstatistics* by Christian Beck and Ezechiel Godert David Cohen in 2003.

This theory is a superposition of two or more different statistics. One is given by the ordinary Boltzmann factor, and the other is given by large-scale fluctuations of one or several intensive parameters. These are indicated by the variable β, which can represent the following: the inverse of the temperature; a chemical potential, in a system with inhomogeneous contractions; uniform effective friction; the amplitude of a disturbing noise; or the energy of local dissipation, as in the case of turbulent flows. The corresponding distributions of stationary probabilities are formed as a convolution of the various statistics. All superstatistics generally have nonadditive and nonextensive entropies.

A special case of superstatistics is nonextensive statistical mechanics, which is formed for the χ^2 distribution; however, other generalized statistics are also possible. To understand the ideas of C. Beck and E. G. D. Cohen, let us consider the distributions $f(\beta)$ and effective Hamiltonians for complex dynamic systems. If E is the energy of a microstate, $B(E)$ can be defined as a generalized Boltzmann factor, for a nonequilibrium system: $B(E) = \int_0^\infty d\beta f(\beta) e^{-\beta E}$. This parameter differs from the usual Boltzmann factor, which is recovered with the Dirac delta distribution $f(\beta) = \delta(\beta - \beta_0)$. It is immediately recognized that $B(E)$ is the Laplace transform of the probability density $f(\beta)$. Other authors, who also work with nonextensive statistical mechanics, interpret $B(E)$ as a convolution of the usual Boltzmann factor with $f(\beta)$.

Consequently, three criteria must be used to obtain the physically important cases: **1.** $f(\beta)$ cannot be any function, it must be a normalized probability density. As an example, we have the Gaussian, uniform, χ^2 and log-normal distributions, etc. However, it can also be some density so far unknown. **2.** The new statistics must be normalized, that is, the integral $\int_0^\infty dE B(E)$ must exist. This is equivalent to saying, in general, that the integral $\int_0^\infty dE \rho(E) B(E)$ must be convergent. $\rho(E)$ represents the density of states. **3.** Any new statistics should be reduced to Boltzmann-Gibbs statistics if it does not have fluctuations in intensive quantities. A series of analytical calculations for different superstatistics—as well as two relevant properties, mean and variance— can be found in the previously cited article by C. Beck and E. G. D. Cohen.

Superstatistics has several applications. Just to cite an example, S. Rizzo and A. Rapisarda (2005) have found that the χ^2 superstatistics describes the behavior of the wind flow velocities at the Florence Airport, Italy. Finally, we must mention that superstatistics can be also applied to the study of econophysics—an interdisciplinary research field, which uses procedures originally developed by physicists. Thereupon, the aim is to solve problems in economics.

Basic Concepts of the Third Version of Nonextensive Statistical Mechanics

Tsallis statistics is part of a plethora of theories that generalize the standard statistical mechanics (also known as Boltzmann-Gibbs-Shannon statistics). For instance, there are theories utilizing, respectively, the Sharma-Mittal, Renyi, Abe, Kaniadakis, and Sharma-Mittal entropies. As a result of its application to multiple and varied systems, nonextensive statistical mechanics has been widely refined since its proposal in 1988. Currently, the enormous number of publications suggests that this statistical mechanics can help us understand some of the anomalous phenomena in statistical physics (Website 1, 2022).

Tsallis Entropy

In 1988, the Greek-Brazilian professor Constantino Tsallis independently again postulated the Havrda-Charvat-Daroczy entropy, with a different constant, and in another context to the one presented by the scientists who first used this entropy. Thus, for a discrete system, we have nonextensive entropy or Tsallis entropy:

$$S_q = k\frac{1-\mathbf{Tr}(\rho^q)}{q-1} \equiv \frac{1-\sum_i p_i^q}{q-1} \ , \qquad (2.1)$$

where ρ^q is the nonextensive density matrix raised to the power q; q, the entropic index; and k, a constant. $\mathbf{Tr}(\rho^q)$ symbolizes the trace of ρ^q and $p_i^q = (\rho_{ii})^q$; likewise, $\mathbf{Tr}(\rho^q) \equiv \sum_i p_i^q$. Besides, if $q \to 1$, Boltzmann-Gibbs-Shannon entropy is recovered—it is also called Boltzmann-Gibbs entropy or von Neumann entropy.

In the context of the third version of statistical mechanics, in applying the maximum entropy method to S_q, for a canonical ensemble, we will need to consider two constraints: unitary norm and nonextensive internal energy, respectively,

$$\mathbf{Tr}(\rho) = 1 \qquad (2.2a)$$

$$U_q = \frac{\mathbf{Tr}\,(H\rho^q)}{\mathbf{Tr}\,(\rho^q)} \ , \qquad (2.2b)$$

where H is the Hamiltonian matrix. Thereupon, after application of the maximum entropy method, we obtain this matrix element:

$$\rho_{ii} \equiv p_i = \frac{1}{Z_q}[1-(1-q)\beta'\varepsilon_i]_+^{\frac{1}{1-q}} \ , \qquad (2.3)$$

obviously, $\rho_{ij} = 0$ when $i \neq j$. Besides, ε_i is the corresponding energy eigenvalue; β', containing the Lagrange multiplier β, is the following energy parameter:

$$\beta' = \frac{\beta}{\mathbf{Tr}(\rho^q) + (1-q)\beta U_q} \ , \qquad (2.4)$$

likewise, Z_q is the generalized partition function:

$$Z_q = \mathbf{Tr}[\mathbb{1} - (1-q)\beta' H]_+^{\frac{1}{1-q}} \ , \qquad (2.5)$$

where $\mathbb{1}$ is the identity matrix; it must be also emphasized that the symbol $[\ \ ]_+$ represents the Tsallis cutoff, that is, the imposition that

$$[1-(1-q)\beta'\varepsilon_i]_+ = \begin{cases} 1-(1-q)\beta'\varepsilon_i \ , & \text{if } 1-(1-q)\beta'\varepsilon_i \geq 0 \\ 0, & \text{else} \end{cases} \ , \qquad (2.6)$$

obviously, this condition ensures that populations will be positive.

For a system composed of two parts A and B, the total entropy of the system A + B is postulated this way:

$$S_{\mathrm{A+B}} = k\frac{1-\mathbf{Tr}_{(\mathrm{A,B})}(\rho^q)}{q-1}\ . \tag{2.7}$$

Similarly, the subsystem entropies are postulated as follows:

$$\mathrm{S_A} = k\frac{1-\mathbf{Tr}_{\mathrm{A}}(\rho_{\mathrm{A}}^q)}{q-1} \quad \text{and} \quad \mathrm{S_B} = k\frac{1-\mathbf{Tr}_{\mathrm{B}}(\rho_{\mathrm{B}}^q)}{q-1}\ , \tag{2.8}$$

where we have the following partial density matrices:

$$\rho_{\mathrm{A}} = \mathbf{Tr}_{\mathrm{B}}(\rho) \text{ and } \quad \rho_{\mathrm{B}} = \mathbf{Tr}_{\mathrm{A}}(\rho)\ , \tag{2.9}$$

these ones are also known as marginal density matrices or reduced density matrices. Obviously, in Eq. (2.8), they appear raised to the power q, that is,

$$\rho_{\mathrm{A}}^q = [\mathbf{Tr}_{\mathrm{B}}(\rho)]^q \quad \text{and} \quad \rho_{\mathrm{B}}^q = [\mathbf{Tr}_{\mathrm{A}}(\rho)]^q\ . \tag{2.10}$$

It should be also mentioned that entropy S_q satisfies the usual properties of positivity, equiprobability, concavity, and irreversibility; however, it does not satisfy the additive property of a system composed of statistically independent subsystems.

Concerning the density matrix ρ, Eq. (2.3)-type distributions can be considered, but with continuous variables, as a special type of stellar polytropes. In astronomy, polytropes represent a class of solutions to the equations of stellar structure. Specifically, the aforementioned distributions are stationary solutions of the Vlasov equation; these distributions were introduced by H. C. Plummer in 1911 and by A. S. Eddington in 1916. Thus, stellar polytropes can be studied in the context of nonextensive statistical mechanics.

Demonstrating How to Obtain Tsallis Entropy Without Postulating It

On the one hand, firstly, we bear in mind that Boltzmann-Gibbs-Shannon entropy is given as follows:

$$S_{\mathrm{BGS}} = -k_{\mathrm{B}}\sum_i\left[p_i^{\mathrm{BGS}}\mathrm{Ln}(p_i^{\mathrm{BGS}})\right]\ , \tag{2.11}$$

where the probability mass function p_i has the following expression:

$$p_i^{\mathrm{BGS}} = \frac{\exp(-\beta\varepsilon_i)}{\sum_j \exp(-\beta\varepsilon_j)}; \tag{2.12}$$

ε_i are the corresponding energy eigenvalue. Then, S_{BGS} can be also expressed this way:

$$S_{\mathrm{BGS}} = -k_{\mathrm{B}}\sum_i\left\{\frac{\exp(-\beta\varepsilon_i)}{\sum_j \exp(-\beta\varepsilon_j)}\mathrm{Ln}\left[\frac{\exp(-\beta\varepsilon_i)}{\sum_j \exp(-\beta\varepsilon_j)}\right]\right\}; \tag{2.13}$$

secondly, we also remember that Tsallis entropy was postulated thus:

$$S_q = k\frac{1-\sum_i p_i^q}{q-1}, \qquad (2.14)$$

with the probability mass function $p_i = \frac{[1-(1-q)\beta'\varepsilon_i]_+^{\frac{1}{1-q}}}{Z_q}$; β' is the energy parameter given by $\beta' = \frac{\beta}{\sum_i p_i^q + (1-q)\beta U q}$. Besides, β is the Lagrange multiplier. Likewise, Z_q is the generalized partition function expressed as $Z_q = \sum_j [1-(1-q)\beta'\varepsilon_j]_+^{\frac{1}{1-q}}$.

On the other hand, we will consider the following two mathematical formulae for the irrational and transcendental number known as Euler's number or Napier's constant:

$$\mathrm{e} = \lim_{n\to\infty}\left(1+\frac{1}{n}\right)^n, \qquad (2.15a)$$

$$\mathrm{e} = \lim_{m\to 0}(1+m)^{\frac{1}{m}}, \qquad (2.15b)$$

to obtain Eq. (2.15b), we have considered the following change of variable $m = 1/n$. Next, we will make other change of variable: $q = 1-m$. Thus, Euler's number is expressed this way:

$$\mathrm{e} = \lim_{q\to 1}[1+(1-q)]^{\frac{1}{1-q}}, \qquad (2.16)$$

now, we can consider the well-known formula for the exponential function:

$$\exp(x) = \mathrm{e}^x = \lim_{n\to\infty}\left(1+\frac{x}{n}\right)^n, \qquad (2.17)$$

which can also be expressed as follows:

$$\exp(x) = \lim_{q\to 1}[1+(1-q)x]^{\frac{1}{1-q}}, \qquad (2.18)$$

obviously, after considering $-x$, the following expression is also valid:

$$\exp(-x) = \lim_{q\to 1}[1-(1-q)x]^{\frac{1}{1-q}}. \qquad (2.19)$$

Besides, we know there is a similar formula for the natural logarithm function:

$$\mathrm{Ln}(x) = \lim_{n\to\infty} n(x^{\frac{1}{n}}-1), \qquad (2.20)$$

we can make the change of variable $m = \frac{1}{n}$, and the logarithm is expressed thus:

$$\mathrm{Ln}(x) = \lim_{m\to 0}\frac{1}{m}(x^m-1); \qquad (2.21)$$

also, we will introduce the new variable $q = m+1$, and the logarithm function takes this way:

$$\mathrm{Ln}(x) = \lim_{q\to 1}\frac{1}{q-1}(x^{q-1}-1); \qquad (2.22)$$

replacing this last equation in Eq. (2.11) with $x \equiv (p_i^{\mathrm{BGS}})$, we have the following expressions:

$$S_{\mathrm{BGS}} = -k_{\mathrm{B}} \sum_i [p_i^{\mathrm{BGS}} \mathrm{Ln}(p_i^{\mathrm{BGS}})] = -k_{\mathrm{B}} \sum_i \left\{ p_i^{\mathrm{BGS}} \mathbf{Lim}_{q \to 1} [\tfrac{1}{q-1} [(p_i^{\mathrm{BGS}})^{q-1} - 1]] \right\}. \quad (2.23)$$

Now, we apply the properties of limit, and it follows that Boltzmann-Gibbs-Shannon entropy is given by

$$S_{\mathrm{BGS}} = \mathbf{Lim}_{q \to 1} \left[k_{\mathrm{B}} \frac{1}{1-q} \sum_i [(p_i^{\mathrm{BGS}})^q - p_i^{\mathrm{BGS}}] \right]; \quad (2.24)$$

by using the normalization condition $\sum_i p_i^{\mathrm{BGS}} = 1$ and by simplifying terms, we have that S_{BGS} is stated thus:

$$S_{\mathrm{BGS}} = \mathbf{Lim}_{q \to 1} \left[k_B \frac{1}{1-q} [\sum_i (p_i^{\mathrm{BGS}})^q - 1] \right]. \quad (2.25)$$

Therefore, we can *almost* identify Tsallis entropy S_q inside the external brackets; however, p_i^{BGS} is the probability mass function associated with Boltzmann-Gibbs-Shannon entropy yet. Then, we can express p_i^{BGS} this way:

$$p_i^{\mathrm{BGS}} = \frac{\exp(-\beta \varepsilon_i)}{\sum_j \exp(-\beta \varepsilon_j)} = \frac{\mathbf{Lim}_{r \to 1} \left\{ [1-(1-r)\beta \varepsilon_i]^{\frac{1}{1-r}} \right\}}{\sum_j \mathbf{Lim}_{r \to 1} \left\{ [1-(1-r)\beta \varepsilon_j]^{\frac{1}{1-r}} \right\}}; \quad (2.26)$$

by interchanging limit and summation in the denominator, and by applying the property about the quotient of two limits, we obtain as follows:

$$p_i^{\mathrm{BGS}} = \frac{\mathbf{Lim}_{r \to 1} \left\{ [1-(1-r)\beta \varepsilon_i]^{\frac{1}{1-r}} \right\}}{\mathbf{Lim}_{r \to 1} \left\{ \sum_j [1-(1-r)\beta \varepsilon_j]^{\frac{1}{1-r}} \right\}} = \mathbf{Lim}_{r \to 1} \left\{ \frac{[1-(1-r)\beta \varepsilon_i]^{\frac{1}{1-r}}}{\sum_j [1-(1-r)\beta \varepsilon_j]^{\frac{1}{1-r}}} \right\} = \mathbf{Lim}_{r \to 1} p_i, \quad (2.27)$$

where we have defined the new probability mass function p_i thus:

$$p_i = \frac{[1-(1-r)\beta \varepsilon_i]^{\frac{1}{1-r}}}{\sum_j [1-(1-r)\beta \varepsilon_j]^{\frac{1}{1-r}}}. \quad (2.28)$$

Therefore, we can replace the result of (2.27) into Eq. (2.25) and, consequently, we obtain that S_{BGS} is given by

$$S_{\mathrm{BGS}} = \mathbf{Lim}_{q \to 1} \left[k_B \frac{1}{1-q} \left[\sum_i (p_i^{\mathrm{BGS}})^q - 1 \right] \right] = \mathbf{Lim}_{q \to 1} \left\{ k_B \frac{1}{1-q} \left[\sum_i \left[\mathbf{Lim}_{r \to 1} p_i \right]^q - 1 \right] \right\}, \quad (2.29)$$

by using the power rule of limits, in the inner limit, we have this way:

$$S_{\mathrm{BGS}} = \mathbf{Lim}_{q \to 1} \left[k_B \frac{1}{1-q} \left(\sum_i \mathbf{Lim}_{r \to 1} p_i^q - 1 \right) \right]; \quad (2.30)$$

now, inside the parentheses, after interchanging limit and summation, it follows that

$$S_{\mathrm{BGS}} = \mathbf{Lim}_{q \to 1} \left[k_B \frac{1}{1-q} \left[\mathbf{Lim}_{r \to 1} \left(\sum_i p_i^q \right) - 1 \right] \right], \quad (2.31)$$

next, considering that the limit of a constant is the same constant, we can put the inner limit outside the inner brackets, that is,

$$S_{\text{BGS}} = \underset{q\to1}{\textbf{Lim}}\left[\underset{r\to1}{\textbf{Lim}}\left[k_B\frac{1}{1-q}\left(\textstyle\sum_i p_i^q - 1\right)\right]\right]; \qquad (2.32)$$

now, in this equation, we want to point out that we can combine the two limits into one limit only. Besides, we can introduce the new parameter k satisfying $\underset{q\to1}{\textbf{Lim}}\, k = k_B$. Therefore, the last equation can be expressed thus:

$$S_{\text{BGS}} = \underset{q\to1}{\textbf{Lim}}\left[k\frac{1}{q-1}\left(1-\textstyle\sum_i p_i^q\right)\right], \qquad (2.33)$$

where we can already identify Tsallis entropy, that is,

$$S_{\text{BGS}} = \underset{q\to1}{\textbf{Lim}}\, S_q = \underset{q\to1}{\textbf{Lim}}\left[k\frac{1}{q-1}\left(1-\textstyle\sum_i p_i^q\right)\right], \qquad (2.34)$$

with the probability mass function for Tsallis entropy expressed by using the entropic index q, namely,

$$p_i = \frac{[1-(1-q)\beta\varepsilon_i]^{\frac{1}{1-q}}}{\sum_j[1-(1-q)\beta\varepsilon_j]^{\frac{1}{1-q}}}\ . \qquad (2.35)$$

In short, we have obtained Tsallis entropy without the need to postulate it.

Demonstrating That the q-Exponential and q-Logarithm Are Inverse Functions

Firstly, we must point out that, within the framework of nonextensive statistical mechanics, there is the q-algebra, which is a set of new mathematical operations and expressions used in that statistical theory (in Appendices B and C, we will demonstrate several additional properties concerning the q-algebra).

Thus, we must start by defining two expressions previously used. This time we will call them the q-exponential and q-logarithm functions, respectively,

$$\exp_q(x) = \begin{cases}[1+(1-q)\,x]^{\frac{1}{1-q}}, & \text{if } [1+(1-q)x] > 0 \text{ and } x\in\mathbb{R} \\ 0, & \text{otherwise}\end{cases} \qquad (2.36)$$

and

$$\text{Ln}_q(y) = \frac{y^{1-q}-1}{1-q}, (y \geq 0,\ q\in\mathbb{R})\ . \qquad (2.37)$$

Likewise, for $q = 1$, we will recover the known mathematical formulae:

$$\exp(x) = \underset{q\to1}{\textbf{Lim}}\exp_q(x) \text{ and } \quad \text{Ln}(y) = \underset{q\to1}{\textbf{Lim}}\,\text{Ln}_q(y)\ . \qquad (2.38)$$

As for the demonstration we are looking for, in the definition of the q-exponential function, Eq. (2.36), we will consider $x = \text{Ln}_q(y)$. Then, we have

$$\exp_q\left(\text{Ln}_q(y)\right) = [1 + (1 - q)\,\text{Ln}_q(y)]^{\frac{1}{1-q}}\,; \qquad (2.39)$$

now, we will replace the respective expression for $\text{Ln}_q(y)$,

$$\exp_q\left(\text{Ln}_q(y)\right) = \left\{1 + (1 - q)\left[\frac{y^{1-q} - 1}{1 - q}\right]\right\}^{\frac{1}{1-q}}, \qquad (2.40)$$

after cross canceling the factor $(1 - q)$, we obtain this formula:

$$\exp_q\left(\text{Ln}_q(y)\right) = [1 + y^{1-q} - 1]^{\frac{1}{1-q}} = [y^{1-q}]^{\frac{1}{1-q}}, \qquad (2.41)$$

and after using the power of a power property, we get $\exp_q\left(\text{Ln}_q(y)\right) = y$.

This way, it has been demonstrated that certainly the q-exponential function $\exp_q(x)$ and the q-logarithm function $\text{Ln}_q(x)$ are inverse functions.

On the other hand, for $q = 1$, we will apply the previous results $\exp(x) = \mathbf{Lim}_{q\to 1} \exp_q(x)$ and $\text{Ln}(y) = \mathbf{Lim}_{q\to 1} \text{Ln}_q(y)$. Therefore, it is valid

$$\exp_1(\text{Ln}_1(y)) = \exp(\text{Ln}(y)) = y\,. \qquad (2.42)$$

which is the expected result.

The Connection Between Renyi Entropy and Tsallis Entropy

These entropies have the same probability distribution. Renyi entropy is defined by the following formula:

$$S^{\text{R}} = k\frac{\text{Ln}\left(\sum_i^N p_i^q\right)}{1 - q}, \qquad (2.43)$$

where 'R' means Renyi, k is a constant, and N stands for the total number of accessible states of the system. Besides, when we deal with a system A + B, with A and B statistically independent, it is formulated as follows:

$$S_{\text{A+B}}^{\text{R}} = S_{\text{A}}^{\text{R}} + S_{\text{B}}^{\text{R}}\,. \qquad (2.44)$$

On the other hand, from Tsallis entropy $S_q = k\frac{1}{q-1}\left(1 - \sum_i^N p_i^q\right)$, we solve for $\sum_i^N p_i^q$. Thus $\sum_i^N p_i^q = 1 + \frac{(1-q)}{k} S_q$. Replacing this expression in the definition of Renyi entropy, Eq. (2.43), we have that Renyi entropy and Tsallis entropy are linked by the following relationship:

$$S^{\text{R}} = k\frac{1}{q-1}\text{Ln}\left[1 + \frac{(1-q)}{k}S_q\right] \quad \text{or} \quad \frac{q-1}{k}S_q + \exp\left[\frac{q-1}{k}S^{\text{R}}\right] = 1\,. \qquad (2.45)$$

Furthermore, we must point out that it is also possible to develop a thermodynamic formalism from Renyi entropy.

Pseudo-additivity for Tsallis Entropy

Given a compound system A + B, assuming the existence of statistical independence for subsystems A and B, we have the entropic formula of pseudo-additivity for Tsallis entropy:

$$S_q^{\mathrm{A+B}} = S_q^{\mathrm{A}} + S_q^{\mathrm{B}} + \frac{1}{k}(1-q)S_q^{\mathrm{A}}S_q^{\mathrm{B}}, \qquad (2.46)$$

where $S_q^{\mathrm{A+B}}$ is the entropy of the compound system A + B; $S_{\mathrm{q}}^{\mathrm{A}}$ and $S_{\mathrm{q}}^{\mathrm{B}}$, the corresponding subsystem entropies. The previously mentioned statistical independence implies the following expressions:

$$\rho^q = \rho_{\mathrm{A}}^q \otimes \rho_{\mathrm{B}}^q \quad \text{or} \quad \mathbf{Tr}_{(\mathrm{A,B})}(\rho^q) = \mathrm{Tr}_{\mathrm{A}}[\rho_{\mathrm{A}}^q]\mathrm{Tr}_{\mathrm{B}}[\rho_{\mathrm{B}}^q], \qquad (2.47)$$

where the symbol $\otimes$ stands for the tensor product.

Next, concisely, it is displayed how you can get pseudo-additivity. From the definition of Tsallis entropy for a compound system A + B, Eq. (2.7), we have

$$\frac{q-1}{k}S_q^{\mathrm{A+B}} = 1 - \mathbf{Tr}_{(\mathrm{A,B})}(\rho^q) = 1 - xy, \qquad (2.48)$$

where we have introduced the parameters $x = \mathbf{Tr}_{\mathrm{A}}[\rho_{\mathrm{A}}^q]$ and $\mathrm{y} = \mathbf{Tr}_{\mathrm{B}}[\rho_{\mathrm{B}}^q]$. By performing algebraic operations on Eq. (2.48), we have

$$\frac{q-1}{k}S_q^{\mathrm{A+B}} = 1 - (-x+1-1)(-y+1-1), \qquad (2.49)$$

then,

$$\frac{q-1}{k}S_q^{\mathrm{A+B}} = 1 - (1-x)(1-y) + (1-x) + (1-y) - 1, \quad (2.50)$$

by replacing $\frac{q-1}{k}S_q^{\mathrm{A}} = 1 - x$ and $\frac{q-1}{k}S_q^{\mathrm{B}} = 1 - y$, we obtain

$$\frac{q-1}{k}S_q^{\mathrm{A+B}} = -\frac{q-1}{k}S_q^{\mathrm{A}}\frac{q-1}{k}S_q^{\mathrm{B}} + \frac{q-1}{k}S_q^{\mathrm{A}} + \frac{q-1}{k}S_q^{\mathrm{B}}, \quad (2.51)$$

after simplifying, we recover Eq. (2.46). Analogously, we have the pseudo-additive formula for three statistically independent systems (A, B, and C),

$$S_q^{\mathrm{A+B+C}} = S_q^{\mathrm{A}} + S_q^{\mathrm{B}} + S_q^{\mathrm{C}} + \frac{1}{k}(1-q)\left(S_q^{\mathrm{A}}S_q^{\mathrm{B}} + S_q^{\mathrm{A}}S_q^{\mathrm{C}} + S_q^{\mathrm{B}}S_q^{\mathrm{C}}\right) + \frac{1}{k^2}(1-q)^2\left(S_q^{\mathrm{A}}S_q^{\mathrm{B}}S_q^{\mathrm{C}}\right). \quad (2.52)$$

On the other hand, there is also normalized Tsallis entropy S_q^{N}, i.e., normalized S_q ('N' means normalized). This entropy was proposed by the Indian A. K. Rajagopal and the Japanese S. Abe in 1999. Thus, for a compound system A + B, we have

$$S_q^{\mathrm{N}} = S_{\mathrm{N}}^{\mathrm{A+B}} = k\frac{1-\mathbf{Tr}_{(\mathrm{A,B})}(\rho^q)}{(q-1)\mathbf{Tr}_{(\mathrm{A,B})}(\rho^q)}; \qquad (2.53)$$

in this context, the entropic index q is interpreted as a measurement concerning the system's degree of extensiveness. Also, it is possible to obtain q from first principles, that is, we can determine this parameter from microscopic and mesoscopic dynamics. Besides, the entropic index q provides a measure of the temperature fluctuations in a system given: $q = \frac{\langle\beta^2\rangle}{\langle\beta\rangle^2}$, where $\langle\beta\rangle$ is the average value of β. Concerning β, it may be different from the temperature. However, this is valid for many statistics apart from nonextensive statistical mechanics. If fluctuations do not occur, Boltzmann-Gibbs statistical mechanics is recovered.

The total system entropy, S_q, can be subextensive, extensive, or super extensive, depending on whether $q > 1$, $q = 1$, or $q < 1$, respectively. The first case favors rare events, i.e., when $p_i^q < p_i$; the second case corresponds to the standard one; the third case favors frequent events, i.e., when $p_i^q > p_i$. The appearance of the factor k in the pseudo-additivity is a matter of discussion; some researchers consider that it is relevant to physical interpretations. However, others warn that the pseudo-additivity is a relationship for thermodynamic quantities; therefore, k should not be included in the pseudo-additivity, analogously to the Boltzmann constant. This one is an intrinsic feature of statistical mechanics and never appears in fundamental thermodynamic relationships.

Definition of Temperature

In nonextensive statistical mechanics, we have several possible definitions of temperature; next, we will analyze three of them.

The first definition originates from the Lagrange multiplier β; this parameter is considered to be the temperature of the system; for non-equilibrium systems, the proposal is to consider Tsallis entropy as Clausius thermodynamic entropy, whenever the Lagrange multiplier β is the inverse temperature. Thus, we have the following relationship:

$$T = \frac{1}{\beta} \,. \qquad (2.54)$$

The second definition of temperature considers the temperature T^*, usually called 'physical' temperature; this parameter is given by

$$T^* = \frac{1}{k\beta^*} = T_{\text{phys.}} = \left[1 + \frac{1-q}{k} S_q\right] \left\{\frac{\partial S_q}{\partial U_q}\right\}^{-1} \,. \qquad (2.55)$$

For $q = 1$, T^* becomes the temperature T from Boltzmann-Gibbs statistics (this result applies to all definitions of temperature in this subsection).

Finally, we have the following definition:

$$T' = \frac{1}{k\beta'} \,, \tag{2.56}$$

where β' has already been defined in Eq. (2.4); it is an auxiliary parameter normally used in the scientific literature. In that sense, in the calculations introduced in Chapters 3 and 4, this definition of temperature is used along with $k = k_{\mathrm{B}}$; k_{B} is the Boltzmann constant. Furthermore, a mathematical relationship alternative to Eq. (2.4), $\beta' = \frac{\beta}{\mathrm{Tr}(\rho^q)+(1-q)\beta U_q}$, is the following one:

$$T' = T\left[1 + (1-q)\frac{S_q}{k}\right] + (1-q)\frac{U_q}{k} \,, \tag{2.57}$$

where we assumed the Lagrange multiplier $\beta = \frac{1}{kT}$. Besides, U_q is the nonextensive internal energy defined in Eq. (2.2b).

Nonextensive Quantum Mean Values

For a physical system, within the framework of the third version of nonextensive statistical mechanics, the thermal quantum mean value of an observable, represented by the operator $\hat{O}$, has two equivalent expressions:

$$\langle \hat{O} \rangle = \frac{\mathbf{Tr}(\hat{\rho}^q \hat{O})}{\mathbf{Tr}(\hat{\rho}^q)} \equiv \frac{\mathbf{Tr}(\rho^q O)}{\mathbf{Tr}(\rho^q)} \,, \tag{2.58}$$

where O is the matrix representation of the operator $\hat{O}$; ρ is the matrix representation of the operator $\hat{\rho}$, ρ was already defined in Eq. (2.3). Similarly, we will define the mean values for a compound system $\mathrm{A}+\mathrm{B}$. Thus, the mean value of an observable of the subsystem A, in the full Hilbert space $\mathrm{A} + \mathrm{B}$, is given as follows:

$$\langle \hat{O}_{\mathrm{A}} \rangle = \frac{\mathbf{Tr}(\rho^q O_{\mathrm{A}})}{\mathbf{Tr}(\rho^q)} \,. \tag{2.59}$$

Analogously, for an observable of the subsystem B, we have

$$\langle \hat{O}_{\mathrm{B}} \rangle = \frac{\mathbf{Tr}(\rho^q O_{\mathrm{B}})}{\mathbf{Tr}(\rho^q)} \,, \tag{2.60}$$

where O_{A} and O_{B} are matrices representing, respectively, observables of the subsystems A and B, i.e.,

$$O_{\mathrm{A}} = \mathcal{O}_{\mathrm{A}} \otimes \mathbb{1}_{\mathrm{B}} \quad \text{and} \quad O_{\mathrm{B}} = \mathbb{1}_{\mathrm{A}} \otimes \mathcal{O}_{\mathrm{B}} \,, \tag{2.61}$$

$\mathcal{O}_{\mathrm{A}}$ and $\mathcal{O}_B$ are the matrices for each observable in the Hilbert subspaces, respectively; $\mathbb{1}_{\mathrm{A}}$ and $\mathbb{1}_{\mathrm{B}}$ are the identity matrices in the corresponding Hilbert subspaces. Likewise, we have an additional method from the concept of partial trace in the corresponding Hilbert subspaces:

$$\langle \hat{\mathcal{O}}_{\mathrm{A}} \rangle = \frac{\mathbf{Tr}_{\mathrm{A}}(\rho_{\mathrm{A}}^q \mathcal{O}_{\mathrm{A}})}{\mathbf{Tr}_{\mathrm{A}}(\rho_{\mathrm{A}}^q)} \,, \tag{2.62a}$$

$$\langle \hat{\mathcal{O}}_{\mathrm{B}} \rangle = \frac{\mathbf{Tr}_{\mathrm{B}}\left(\rho_{\mathrm{B}}^{q} \mathcal{O}_{\mathrm{B}}\right)}{\mathbf{Tr}_{\mathrm{B}}\left(\rho_{\mathrm{B}}^{q}\right)}, \qquad (2.62b)$$

where ρ_{A}^{q} and ρ_{B}^{q} were already defined in Eq. (2.10). Therefore, since an operator represents a property of a subsystem, its mean value in the compound-system space must be equal to its mean value in the respective subsystem space, that is, Eq. (2.59) must be equivalent to Eq. (2.62a) (the concept of partial trace guarantees this):

$$\langle \hat{\mathcal{O}}_{\mathrm{A}} \rangle_{\mathrm{FHS}} \equiv \langle \hat{\mathcal{O}}_{\mathrm{A}} \rangle_{\mathrm{HS}} \quad \text{and} \quad \langle \hat{\mathcal{O}}_{\mathrm{B}} \rangle_{\mathrm{FHS}} \equiv \langle \hat{\mathcal{O}}_{\mathrm{B}} \rangle_{\mathrm{HS}}, \qquad (2.63)$$

where FHS means *full Hilbert space* and HS, *Hilbert subspace*. Next, for illustrative purposes, we will show the formulae of nonextensive magnetization and internal energy. These parameters have been chosen because the thermodynamic quantities can be verified experimentally. The magnetization can be measured, for example, in a commercial SQUID magnetometer like the CBPF SQUID. Concerning the internal energy, even though the absolute values of this parameter cannot usually be determined by using calorimetric techniques, it is possible to measure variations of internal energy in a process at constant volume. Thus, the nonextensive magnetization along the *z*-axis is defined as follows:

$$M = M_z = \frac{1}{m} \langle \hat{\mu}_z \rangle = \frac{1}{m} \frac{\mathbf{Tr}(\rho^q \mu_{\mathrm{z}})}{\mathbf{Tr}(\rho^q)}, \qquad (2.64)$$

where $\hat{\mu}_z$ stands for the magnetic dipole moment operator (μ_{z} is the corresponding matrix); m, the system mass; and ρ^q, the density matrix raised to the power q. This dipole moment is defined as $\mu_{\mathrm{z}} = gJ_{\mathrm{z}}\mu_{\mathrm{Bohr}}$, where g is the gyromagnetic factor, also called the Landé factor; J_{z}, the angular momentum matrix (J_{z} is arbitrary and dimensionless); and μ_{Bohr}, the Bohr magneton $\mu_{\mathrm{Bohr}} = 9.274 \times 10^{-24}$ JT^{-1} in the international system of units and $\mu_{\mathrm{Bohr}} = 9.274\mathrm{x}10^{-21}$ emu in electromagnetic units (1 emu=10^{-3} JT^{-1}).

In the case of a compound system A + B, in the full Hilbert space, we have the formulae of the nonextensive magnetization for each subsystem:

$$M_{\mathrm{A}} = \frac{\mathbf{Tr}(g_{\mathrm{A}} \rho^q J_z^{\mathrm{A}})}{\mathbf{Tr}(\rho^q)} \frac{\mu_{\mathrm{Bohr}}}{m_{\mathrm{A}}}, \qquad (2.65a)$$

$$M_{\mathrm{B}} = \frac{\mathbf{Tr}(g_{\mathrm{B}} \rho^q J_z^{\mathrm{B}})}{\mathbf{Tr}(\rho^q)} \frac{\mu_{\mathrm{Bohr}}}{m_{\mathrm{B}}}, \qquad (2.65b)$$

where g_{A} and g_{B} are the Landé factors for each subsystem. Besides, J_z^{A} and J_z^{B} are the angular momentum matrices in the full Hilbert space, that is, $J_z^{\mathrm{A}} = \mathbb{j}_z^{\mathrm{A}} \otimes \mathbb{1}_{\mathrm{B}}$ and $J_z^{\mathrm{B}} = \mathbb{1}_{\mathrm{A}} \otimes \mathbb{j}_z^{\mathrm{B}}$, where $\mathbb{j}_z^{\mathrm{A}}$ and $\mathbb{j}_z^{\mathrm{B}}$ are angular momentum matrices in the respective Hilbert subspaces; $\mathbb{1}_{\mathrm{A}}$ and $\mathbb{1}_{\mathrm{B}}$ are the identity matrices in the respective Hilbert subspaces; and the symbol $\otimes$

stands for the tensor product. All the aforementioned angular momentum matrices are dimensionless.

In the Hilbert subspaces, we have the following nonextensive magnetizations:

$$M_{\mathrm{A}} = \frac{\mathbf{Tr}\left(g_{\mathrm{A}}\rho_{\mathrm{A}}^{q}\mathrm{j}_{z}^{\mathrm{A}}\right)}{\mathbf{Tr}\left(\rho_{\mathrm{A}}^{\mathrm{q}}\right)}\frac{\mu_{\mathrm{Bohr}}}{m_{\mathrm{A}}}, \qquad (2.66a)$$

$$M_{\mathrm{B}} = \frac{\mathbf{Tr}\left(g_{\mathrm{B}}\rho_{\mathrm{B}}^{q}\mathrm{j}_{z}^{\mathrm{B}}\right)}{\mathbf{Tr}\left(\rho_{\mathrm{B}}^{q}\right)}\frac{\mu_{\mathrm{Bohr}}}{m_{\mathrm{B}}}, \qquad (2.66b)$$

where ρ_{A}^{q} and ρ_{B}^{q} are the partial density matrices raised to the power q, which were defined in Eq. (2.10).

On the other hand, in a way similar, we can define the nonextensive internal energies. Then, in the *full* Hilbert space, we have the following energies:

$$U_{\mathrm{A}} = \frac{\mathbf{Tr}(\rho^{q}H_{\mathrm{A}})}{\mathbf{Tr}(\rho^{q})}, \qquad (2.67a)$$

$$U_{\mathrm{B}} = \frac{\mathbf{Tr}(\rho^{q}H_{\mathrm{B}})}{\mathbf{Tr}(\rho^{q})}, \qquad (2.67b)$$

where H_{A} and H_{B} are the Hamiltonian matrices in the full Hilbert space, but concerning only states A and B, respectively.

Finally, in the Hilbert *subspaces*, we have the following nonextensive internal energies:

$$U_{\mathrm{A}} = \frac{\mathbf{Tr}\left(\rho_{\mathrm{A}}^{q}\mathcal{H}_{\mathrm{A}}\right)}{\mathbf{Tr}\left(\rho_{\mathrm{A}}^{q}\right)}, \qquad (2.68a)$$

$$U_{\mathrm{B}} = \frac{\mathbf{Tr}\left(\rho_{\mathrm{B}}^{q}\mathcal{H}_{\mathrm{B}}\right)}{\mathbf{Tr}\left(\rho_{\mathrm{B}}^{q}\right)}, \qquad (2.68b)$$

where $\mathcal{H}_{\mathrm{A}}$and $\mathcal{H}_{\mathrm{B}}$ are the Hamiltonian matrices in the corresponding Hilbert subspaces.

Escort Distributions

For the mean values from Eq. (2.58), an alternative formalism is to use escort distributions. Thus, the mean value of the observable O, represented by the operator $\hat{O}$, is given by

$$\langle\hat{O}\rangle = \mathbf{Tr}(OP), \qquad (2.69)$$

where O and P are matrices. P is an escort distribution, which was created outside the context of nonextensive statistical mechanics. Thus, the escort distribution of ρ, with order q, is given as follows:

$$P = \frac{\rho^q}{\mathbf{Tr}(\rho^q)} \,, \qquad (2.70)$$

likewise, ρ is the escort distribution of P, with order $\frac{1}{q}$:

$$\rho = \frac{P^{\frac{1}{q}}}{\mathbf{Tr}\left(P^{\frac{1}{q}}\right)} \,. \qquad (2.71)$$

Concerning the mean values for a compound system A + B, we have the following formulae:

$$\langle \hat{O}_{\mathrm{A}} \rangle = \mathbf{Tr}\left(P_{\mathrm{A}} \hat{O}_{\mathrm{A}}\right) \quad \text{and} \quad \langle \hat{O}_{\mathrm{B}} \rangle = \mathbf{Tr}\left(P_{\mathrm{B}} \hat{O}_{\mathrm{B}}\right) \,, \qquad (2.72)$$

where $\hat{O}_{\mathrm{A}}$ and $\hat{O}_{\mathrm{B}}$ are operators representing, respectively, an observable in each Hilbert subspace; P_{A} and P_{B} stand for the following partial matrices:

$$P_{\mathrm{A}} = \mathbf{Tr}_{\mathrm{B}}(P) \,, \qquad (2.73a)$$

$$P_{\mathrm{B}} = \mathbf{Tr}_{\mathrm{A}}(P) \,. \qquad (2.73b)$$

Escort Entropy

As an additional proposal for the construction of a thermodynamic formalism, we can use the escort entropy, which is postulated this way:

$$S_q^{\mathrm{E}} = \mathrm{k}\frac{1 - \left[\mathbf{Tr}\left(P^{\frac{1}{q}}\right)\right]^{-q}}{q-1} \,. \qquad (2.74)$$

where the superscript 'E' means escort. This entropic form is obtained from Tsallis entropy, S_q, by making a change of variable and using the escort distributions, too. Following this idea, for a compound system A + B, we can postulate the respective subsystem entropies as follows:

$$S_{\mathrm{A}}^{\mathrm{E}} = \mathrm{k}\frac{1-\left[\mathbf{Tr}_{\mathrm{A}}\left(P_{\mathrm{A}}^{\frac{1}{q}}\right)\right]^{-q}}{q-1} \quad \text{and} \quad S_{\mathrm{B}}^{\mathrm{E}} = \mathrm{k}\frac{1-\left[\mathbf{Tr}_{\mathrm{B}}\left(P_{\mathrm{B}}^{\frac{1}{q}}\right)\right]^{-q}}{q-1} \,, \qquad (2.75)$$

the partial matrices P_{A} and P_{B} have already been defined in Eqs. (2.73).

Therefore, to obtain the probability distribution, we must maximize Eq. (2.74) considering the constraints of unitary norm $\mathbf{Tr}(P) = 1$, and nonextensive internal energy $U_q = \mathbf{Tr}(PH)$. The probability distribution obtained is represented by the following density matrix:

$$P = \frac{\left[\mathbb{1} - (1-q)\overline{\beta}' H\right]_{+}^{\frac{q}{1-q}}}{\bar{Z}_q}\,, \tag{2.76}$$

$\bar{Z}_q$ is the partition function:

$$\bar{Z}_q = \mathbf{Tr}\left\{\left[\mathbb{1} - (1-q)\overline{\beta}' H\right]_{+}^{\frac{q}{1-q}}\right\}, \tag{2.77}$$

with the energy parameter defined in this manner:

$$\overline{\beta}' = \frac{\bar{\beta}}{\left\{\mathbf{Tr}\left(P^{\frac{1}{q}}\right)\right\}^{-q} + (1-q)\bar{\beta}\, U_q}\,, \tag{2.78}$$

where $\bar{\beta}$ is the customary Lagrange multiplier.

Likewise, in the Hilbert subspaces, the nonextensive magnetizations are given by

$$M_{\mathrm{A}} = \mathbf{Tr}(g_{\mathrm{A}} P_{A} \jmath_z^{A}) \frac{\mu_{\mathrm{Bohr}}}{m_{\mathrm{A}}} \quad \text{and} \quad M_{\mathrm{B}} = \mathbf{Tr}(g_{\mathrm{B}} P_{B} \jmath_z^{B}) \frac{\mu_{\mathrm{Bohr}}}{m_{\mathrm{B}}}\,; \tag{2.79}$$

similarly, we define the nonextensive internal energies this way:

$$U_{\mathrm{A}} = \mathbf{Tr}_{\mathrm{A}}(P_{\mathrm{A}} \mathcal{H}_{\mathrm{A}}) \quad \text{and} \quad U_{\mathrm{B}} = \mathbf{Tr}_{\mathrm{B}}(P_{\mathrm{B}} \mathcal{H}_{\mathrm{B}})\,, \tag{2.80}$$

where $\jmath_z^{A}$, $\jmath_z^{B}$, μ_{Bohr}, m_{A}, m_{B}, $\mathcal{H}_A$, and $\mathcal{H}_B$ have already been identified above.

Entropy Optimization: Maximization and Minimization

Following different formulations, Edwin Thompson Jaynes and Solomon Kullback proposed principles of entropy optimization. As the Jaynes principle has already been quickly explained, we must point out that the Kullback principle involves the search for the probability distribution closest to another *a priori* distribution, by minimizing a measure of divergence between the two distributions. Both Shannon's and Kullback's measures are intrinsically nonlinear probability functions. Thus, the Jaynes principle of maximum entropy and the Kullbacks principle of minimum cross-entropy are reduced to non-linear programming problems whose solution requires an iterative search algorithm.

In concluding this section, it is important to mention that there is a relationship between thermodynamic entropy and information entropy. This relationship is illustrated, generally, with the famous *gedankenexperiment* by J. C. Maxwell (*gedankenexperiment* is the German word for thought experiment). In this one, Maxwell's demon (an imaginary being) needs information to decrease thermodynamic entropy. The disposal of this information—that is, the erasure of memory, so that new observations can be made—dissipates energy. The acquisition of information apparently does not cause such dissipation (however, this is still a point of contention).

So, the energy expenditure exactly balances the thermodynamic gain, which the devil would otherwise achieve. This explanation, dominant in the scientific literature, was proposed by the American physicist and cryptologist Charles Henry Bennett in 1987. Examples of real Maxwell's demons happen in biological systems, such as ion bombs that allow our nervous system to function. Even more, some authors consider that a realistic example of Maxwell's Demon is the evolution by natural selection. Thereupon, it is very interesting the work of Ch. Busby and Ch. V. Howard (2017) entitled *Re-Thinking Biology—I. Maxwell's Demon and the Spontaneous Origin of Life*.

Some Demonstrations by Using the Maximum Entropy Method

We will show the procedures to maximize Boltzmann-Gibbs-Shannon entropy and Tsallis entropy. For the former, we have one procedure; for the latter, we have three procedures because there are three versions of nonextensive statistical mechanics. In other words, we will find the value of the respective probability mass function that maximizes the entropies mentioned above.

Boltzmann-Gibbs-Shannon Entropy Maximization

Boltzmann-Gibbs-Shannon Entropy Maximization for a Canonical Ensemble

Firstly, applying the Jaynes methodology, we will use the method of Lagrange multipliers to maximize the following expression (α and β are the Lagrange multipliers):

$$L = S_{\mathrm{BGS}} - \alpha \sum_j p_j - \beta U \, , \quad (2.81)$$

where L is called the Lagrangian function, S_{BGS} stands for Boltzmann-Gibbs-Shannon entropy; U, internal energy; p_j, the probability that the system is in the state j. Besides, the summation index j runs over all accessible states of the system. Likewise, we realize α has the dimension of k_{B} (bearing in mind $S_{\mathrm{BGS}} = k_{\mathrm{B}} \sum_j p_j \mathrm{Ln}(p_j)$) and β, the dimension of inverse temperature.

We also will consider that maximization is subjected to the following constraints of unitary norm and internal energy, respectively,

$$\sum_j p_j = 1 \quad \text{and} \quad U = \sum_j p_j \varepsilon_j \, , \qquad (2.82)$$

where ε_j are the energy autovalues. Now, we will differentiate the Lagrangian function and set it equal to zero:

$$\frac{\partial L}{\partial p_i} = \frac{\partial}{\partial p_i}\left(S_{\text{BGS}} - \alpha \sum_j p_j - \beta U\right) = 0\ , \quad (2.83)$$

replacing values inside the parentheses, we have the following expression:

$$\frac{\partial}{\partial p_i}\left(-k_{\text{B}} \sum_j p_j \text{Ln}(p_j) - \alpha \sum_j p_j - \beta \sum_j p_j \varepsilon_j\right) = 0\ , \quad (2.84)$$

after differentiating, it follows that

$$-k_{\text{B}}\text{Ln}(p_i) - k_{\text{B}} - \alpha - \beta\varepsilon_i = 0\ , \quad (2.85)$$

then, solving for $\text{Ln}(p_i)$, it becomes this way:

$$\text{Ln}(p_i) = -1 - \frac{\alpha}{k_{\text{B}}} - \frac{\beta\varepsilon_i}{k_{\text{B}}}, \quad (2.86)$$

by applying the exponential operation to both sides, p_i is stated in this manner:

$$p_i = \exp\left(-1 - \frac{\alpha}{k_{\text{B}}} - \frac{\beta\varepsilon_i}{k_{\text{B}}}\right), \quad (2.87)$$

by summing over all the accessible states and applying the normalization condition, we obtain the following expression:

$$\sum_i p_i = 1 = \exp\left(-1 - \frac{\alpha}{k_{\text{B}}}\right) \sum_i \exp\left(-\frac{\beta\varepsilon_i}{k_{\text{B}}}\right), \quad (2.88)$$

now, we will solve for $\exp\left(-1 - \frac{\alpha}{k_{\text{B}}}\right)$. Thus, we get the formula

$$\exp\left(-1 - \frac{\alpha}{k_{\text{B}}}\right) = \frac{1}{\sum_i \exp\left(-\frac{\beta\varepsilon_i}{k_{\text{B}}}\right)} = \frac{1}{Z_{\text{c}}}; \quad (2.89)$$

Z_{c} is the partition function (abbreviation 'c' means canonical). Z_{c} is defined as follows:

$$Z_{\text{c}} = \sum_i \exp\left(-\frac{\beta\varepsilon_i}{k_{\text{B}}}\right); \quad (2.90)$$

thus, we can express the probability mass function, Eq. (2.87), this way:

$$p_i = \frac{1}{Z_{\text{c}}} \exp\left(-\frac{\beta\varepsilon_i}{k_{\text{B}}}\right). \quad (2.91)$$

In summary, this probability mass function is the one maximizing Boltzmann-Gibbs-Shannon entropy.

Boltzmann-Gibbs-Shannon Entropy Maximization for a Grandcanonical Ensemble

In this case, Boltzmann-Gibbs-Shannon entropy and the three constraints to be considered are given respectively by the following formulae:

$$S_{\text{BGS}}^{\text{gc}} = -k_{\text{B}} \sum_{j,n_j} p_j \text{Ln}(p_j)\ , \quad (2.92)$$

$$\sum_{j,n_j} p_j = 1,\ U = \sum_{j,n_j} p_j \varepsilon_j,\ \text{ and }\ N = \sum_{j,n_j} p_j n_j\ , \quad (2.93)$$

where the abbreviation 'gc' means grandcanonical. Likewise, p_j stands for the probability mass function for the state j; n_j, number of particles in the state j; U, internal energy; ε_j, energy autovalues; N, mean number of particles. As a next step, we will establish the Lagrangian function this way:

$$L = S_{\text{BGS}} - \alpha \sum_{j,n_j} p_j - \beta U - \gamma N \,. \quad (2.94)$$

where α, β, and γ are the Lagrange multipliers. Besides, we realize α and γ have the dimension of k_{B}; and β, the dimension of inverse temperature.

Now, the procedure consists in differentiating the Lagrangian function with respect to p_i and equalizing the result to zero. Before this, we will express the Lagrangian function as a functional of p_i. For this objective, in Eq. (2.94), we will replace the respective expressions for S_{BGS}, U, and N.

Thus, we obtain as follows:

$$L = -k_{\text{B}} \sum_{j,n_j} p_j \text{Ln}(p_j) - \alpha \sum_{j,n_j} p_j - \beta \sum_{j,n_j} p_j \varepsilon_j - \gamma \sum_{j,n_j} p_j n_j, \quad (2.95)$$

by applying the properties of summation, we get

$$L = \sum_{j,n_j} [-k_{\text{B}} p_j \text{Ln}(p_j) - \alpha p_j - -\beta p_j \varepsilon_j - \gamma p_j n_j]; \quad (2.96)$$

now, we will differentiate the Lagrangian function with respect to p_i and set it equal to zero:

$$\frac{\partial L}{\partial p_i} = \frac{\partial}{\partial p_i} \sum_{j,n_j} [-k_{\text{B}} p_j \text{Ln}(p_j) - \alpha p_j - \beta p_j \varepsilon_j - \gamma p_j n_j] = 0 \,; \quad (2.97)$$

next, we will perform the addition considering the summation indices; afterward, we will differentiate all the addends. Consequently, from Eq. (2.97) we arrive at

$$-k_{\text{B}} \text{Ln}(p_i) - k_{\text{B}} - \alpha - \beta \varepsilon_i - \gamma n_i = 0 \,, \quad (2.98)$$

solving for $\text{Ln}(p_i)$, we have as follows:

$$\text{Ln}(p_i) = -1 - \frac{\alpha}{k_{\text{B}}} - \frac{\beta \varepsilon_i + \gamma n_i}{k_{\text{B}}} \,; \quad (2.99)$$

next, we will take the exponential on both sides; thus, we obtain the following expression:

$$p_i = \exp\left(-1 - \frac{\alpha}{k_{\text{B}}} - \frac{\beta \varepsilon_i + \gamma n_i}{k_{\text{B}}}\right), \quad (2.100)$$

so, we realize $\exp\left(-1 - \frac{\alpha}{k_{\text{B}}} - \frac{\beta \varepsilon_i + \gamma n_i}{k_{\text{B}}}\right)$ can be expressed as the product of two exponential functions, that is,

$$p_i = \exp\left(-1 - \frac{\alpha}{k_{\text{B}}}\right) \exp\left(-\frac{\beta \varepsilon_i + \gamma n_i}{k_{\text{B}}}\right) \,, \quad (2.101)$$

by summing over all the states and by considering the normalization condition, it follows that

$$\sum_{i,n_i} p_i = 1 = \exp\left(-1 - \frac{\alpha}{k_{\text{B}}}\right) \sum_{i,n_i} \exp\left(-\frac{\beta \varepsilon_i + \gamma n_i}{k_{\text{B}}}\right), \quad (2.102)$$

then solving for $\exp\left(-1 - \frac{\alpha}{k_{\text{B}}}\right)$, we obtain this way

$$\exp\left(-1 - \frac{\alpha}{k_{\text{B}}}\right) = \frac{1}{\sum_{i,n_i} \exp\left(-\frac{\beta \varepsilon_i + \gamma n_i}{k_{\text{B}}}\right)} = \frac{1}{Z_{\text{gc}}}, \quad (2.103)$$

where we have introduced the following grandcanonical partition function:

$$Z_{\text{gc}} = \sum_{i,n_i} \exp\left(-\frac{\beta\varepsilon_i + \gamma n_i}{k_{\text{B}}}\right); \qquad (2.104)$$

finally, after replacing Eq. (2.103) into Eq. (2.101), the probability mass function is given as follows:

$$p_i = \frac{\exp\left(-\frac{\beta\varepsilon_i + \gamma n_i}{k_{\text{B}}}\right)}{\sum_{i,n_i} \exp\left(-\frac{\beta\varepsilon_i + \gamma n_i}{k_{\text{B}}}\right)} = \frac{1}{Z_{\text{gc}}} \exp\left(-\frac{\beta\varepsilon_i + \gamma n_i}{k_{\text{B}}}\right). \qquad (2.105)$$

Furthermore, we must point out that the grandcanonical partition function can be expressed this way:

$$Z_{\text{gc}} = \sum_{i,n_i} \exp\left(-\frac{\beta\varepsilon_i + \gamma n_i}{k_{\text{B}}}\right) = \prod_i Z_i\,, \qquad (2.106)$$

where $\prod_{\text{i}} \text{Z}_{\text{i}}$ means the product of all Z_i, which are defined this way:

$$Z_i = \sum_{n_i} \exp\left(-\frac{\beta\varepsilon_i + \gamma n_i}{k_{\text{B}}}\right). \qquad (2.107)$$

Tsallis Entropy Maximization in the First Version of Nonextensive Statistical Mechanics

We will deal with a canonical ensemble. Then, to maximize Tsallis entropy $S_q = k(1 - \sum_i p_i^q)/(q-1)$, we will use the method of the Lagrange multipliers.

First of all, we will form the following Lagrangian function (α and β are Lagrange multipliers):

$$L = S_q - \alpha \sum_j p_j - \beta U_q^{(1)}, \qquad (2.108)$$

where p_j is the probability mass function for the state j; and $U_q^{(1)}$, the internal energy in the framework of the first version of nonextensive statistical mechanics. Likewise, we must consider the constraints of normalization condition and internal energy, respectively:

$$\sum_j p_j = 1 \quad \text{and} \quad U_q^{(1)} = \sum_j \varepsilon_j p_j\,. \qquad (2.109)$$

Next, we partially differentiate the Lagrangian function with respect to p_i and set it equal to zero:

$$\frac{\partial L}{\partial p_i} = \frac{\partial}{\partial p_i}\left(S_q - \alpha \sum_j p_j - \beta U_q^{(1)}\right) = 0\,; \qquad (2.110)$$

afterward, we obtain the following result:

$$\frac{kq}{1-q} p_i^{q-1} - \alpha - \beta\varepsilon_i = 0\,, \qquad (2.111)$$

solving for p_i, it follows that

$$p_i = \left\{\frac{1-q}{kq}(\alpha + \beta\varepsilon_i)\right\}^{\frac{1}{q-1}}; \qquad (2.112)$$

now, we need to know the value of α (we will utilize a similar procedure to obtain α in the other two versions of nonextensive statistical mechanics). From Eq. (2.111), we solve for α and obtain

$$\alpha = \frac{kq}{1-q} p_i^{q-1} - \beta \varepsilon_i \; ; \quad (2.113)$$

next, we multiply Eq. (2.113) by p_i and sum over all states i (considering $\sum_i p_i = 1$). The expression obtained is given as follows:

$$\alpha = \frac{kq}{1-q} \sum_i p_i^q - \beta U_q^{(1)} \, , \qquad (2.114)$$

besides, if we introduce a new parameter named $X_q = \sum_i p_i^q$, the last equation is expressed thus:

$$\alpha = \frac{kq}{1-q} X_q - \beta U_q^{(1)} \; ; \qquad (2.115)$$

then, we replace this value of α into Eq. (2.112), obtaining the following probability mass function:

$$p_i = \left\{ X_q + \frac{1-q}{kq} \beta (\varepsilon_i - U_q^{(1)}) \right\}^{\frac{1}{q-1}} . \qquad (2.116)$$

or, after the corresponding multiplications,

$$p_i = \left\{ X_q + \frac{1-q}{kq} \beta \varepsilon_i - \frac{1-q}{kq} \beta U_q^{(1)} \right\}^{\frac{1}{q-1}} = \left\{ X_q - \frac{1-q}{kq} \beta U_q^{(1)} + \frac{1-q}{kq} \beta \varepsilon_i \right\}^{\frac{1}{q-1}}, \; (2.117)$$

where we have used the commutative property of addition. Now, we will factor out $\left[X_q - \frac{1-q}{kq} \beta U_q^{(1)} \right]$; consequently, p_i is stated as follows:

$$p_i = \left\{ \left[X_q - \frac{1-q}{kq} \beta U_q^{(1)} \right] \left[1 + \frac{1}{X_q - \frac{1-q}{kq} \beta U_q^{(1)}} \frac{1-q}{kq} \beta \varepsilon_i \right] \right\}^{\frac{1}{q-1}} \; ; (2.118)$$

introducing a new (inverse) energy parameter named

$$\beta' = \frac{1}{X_q - \frac{1-q}{kq} \beta U_q^{(1)}} \frac{1}{kq} \beta = \frac{1}{kq X_q - \beta (1-q) U_q^{(1)}} \beta \, , \qquad (2.119)$$

the probability mass function p_i is given by

$$p_i = \left[X_q - \frac{1-q}{kq} \beta U_q^{(1)} \right] [1 + (1-q) \beta' \varepsilon_i]^{\frac{1}{q-1}} \; ; \qquad (2.120)$$

next, we will sum over all states i, taking into account the normalization condition $\sum_i p_i = 1$:

$$1 = \left[X_q - \frac{1-q}{kq} \beta U_q^{(1)} \right] \sum_i [1 + (1-q) \beta' \varepsilon_i]^{\frac{1}{q-1}} \; ; \quad (2.121)$$

therefore, we can define the nonextensive partition function this way:

$$Z_q = \frac{1}{X_q - \frac{1-q}{kq} \beta U_q^{(1)}} = \sum_i [1 + (1-q) \beta' \varepsilon_i]^{\frac{1}{q-1}} \; ; \; (2.122)$$

replacing Eq. (2.122) into Eq. (2.120), we obtain the expression for the probability mass function in the first version of nonextensive statistical mechanics:

$$p_i = \frac{1}{Z_q}[1 + (1-q)\beta' \varepsilon_i]^{\frac{1}{q-1}} \; . \qquad (2.123)$$

This probability mass function is a power-law one.

Tsallis Entropy Maximization in the Second Version of Nonextensive Statistical Mechanics

We will deal with a canonical ensemble. So, we must point out that the procedure for maximizing Tsallis entropy is like the one applied in the first version of nonextensive statistical mechanics; however, the calculations are different. Then, we must again find the probability mass function that maximizes Tsallis entropy subjected to constraints of normalization condition and internal energy. Therefore, firstly, we will form the following Lagrangian function:

$$L = S_q - \alpha \sum_j p_j - \beta U_q^{(2)}, \qquad (2.124)$$

where $S_q = k(1 - \sum_i p_i^q)/(q-1)$; α and β are the Lagrange multipliers; p_j is the probability mass function for the j-th state; and $U_q^{(2)}$ is the internal energy in the framework of the second version of nonextensive statistical mechanics. We must also consider the aforementioned constraints, namely,

$$\sum_j p_j = 1 \quad \text{and} \quad U_q^{(2)} = \sum_j p_j^q \varepsilon_j \; . \qquad (2.125)$$

Consequently, we partially differentiate the Lagrangian function with respect to p_i and set it equal to zero, that is,

$$\frac{\partial L}{\partial p_i} = \frac{\partial}{\partial p_i}\left(S_q - \alpha \sum_j p_j - \beta U_q^{(2)}\right) = 0 \; , \qquad (2.126)$$

after deriving the corresponding expressions, we obtain this way:

$$\frac{kq}{1-q} p_i^{q-1} - \alpha - \beta q \varepsilon_i p_i^{q-1} = 0 \; , \qquad (2.127)$$

solving for p_i^{q-1}, it follows that

$$p_i^{q-1} = \frac{\alpha}{\frac{kq}{1-q} - \beta q \varepsilon_i} \; ; \qquad (2.128)$$

however, we need to know α. Thus, we will multiply the last equation by p_i and sum over all states i (considering the normalization condition $\sum_i p_i = 1$). Afterward, we solve for α, that is,

$$\alpha = \sum_i p_i^q \left[\frac{kq}{1-q} - \beta q \varepsilon_i\right] = \frac{kq}{1-q} X_q - \beta q U_q^{(2)} \; , \qquad (2.129)$$

where we have introduced the parameter $X_q = \sum_i p_i^q$. Next, we will replace Eq. (2.129) into Eq. (2.128); thus, we obtain the following expression:

$$p_i^{q-1} = \frac{\frac{kq}{1-q}X_q - \beta q U_q^{(2)}}{\frac{kq}{1-q} - \beta q \varepsilon_i} = \frac{X_q - \frac{(1-q)\beta U_q^{(2)}}{k}}{1 - \frac{(1-q)\beta\varepsilon_i}{k}}, \qquad (2.130)$$

by raising to the power -1, the previous formula is given as follows:

$$p_i^{1-q} = \frac{1 - \frac{(1-q)\beta\varepsilon_i}{k}}{X_q - \frac{(1-q)\beta U_q^{(2)}}{k}}, \qquad (2.131)$$

thus, we get the following probability mass function:

$$p_i = \left\{\frac{1 - \frac{(1-q)\beta\varepsilon_i}{k}}{X_q - \frac{(1-q)\beta U_q^{(2)}}{k}}\right\}^{\frac{1}{1-q}}. \qquad (2.132)$$

To find the partition function, on both sides we will sum over all accessible states and, afterward, we will consider the normalization condition $\sum_i p_i = 1$. Thus, we obtain

$$1 = \sum_i \left\{\frac{1 - \frac{(1-q)\beta\varepsilon_i}{k}}{X_q - \frac{(1-q)\beta U_q^{(2)}}{k}}\right\}^{\frac{1}{1-q}} = \frac{\sum_i\left[\left\{1 - \frac{(1-q)\beta\varepsilon_i}{k}\right\}^{\frac{1}{1-q}}\right]}{\left\{X_q - \frac{(1-q)\beta U_q^{(2)}}{k}\right\}^{\frac{1}{1-q}}}, \qquad (2.133)$$

then, we can already define the partition function as

$$Z_q = \left\{X_q - \frac{(1-q)\beta U_q^{(2)}}{k}\right\}^{\frac{1}{1-q}} = \sum_i\left[\left\{1 - \frac{(1-q)\beta\varepsilon_i}{k}\right\}^{\frac{1}{1-q}}\right]. \qquad (2.134)$$

Finally, we replace Eq. (2.134) into Eq. (2.132) and obtain the well-known expression for the probability mass function in the second version of nonextensive statistical mechanics:

$$p_i = \frac{\left\{1 - \frac{(1-q)\beta\varepsilon_i}{k}\right\}^{\frac{1}{1-q}}}{\sum_i\left\{1 - \frac{(1-q)\beta\varepsilon_i}{k}\right\}^{\frac{1}{1-q}}} = \frac{1}{Z_q}\left\{1 - \frac{(1-q)\beta\varepsilon_i}{k}\right\}^{\frac{1}{1-q}}, \qquad (2.135)$$

This result is similar to Eq. (2.123) and, evidently, it is a power-law probability mass function.

Tsallis Entropy Maximization in the Third Version of Nonextensive Statistical Mechanics

We will deal with a canonical ensemble. Then, firstly, we must form the Lagrangian function, which is given as follows:

$$L = S_q - \alpha\sum_j p_j - \beta U_q^{(3)}, \qquad (2.136)$$

where $S_q = k(1 - \sum_i p_i^q)/(q-1)$; α and β are the Lagrange multipliers; p_j is the probability mass function for the j-th state; and $U_q^{(3)}$ is the internal energy in the framework of the third version of nonextensive statistical mechanics. Likewise, we will consider two constraints: normalization condition and internal energy; these ones are respectively given by

$$\sum_j p_j = 1 \quad \text{and} \quad U_q^{(3)} = \frac{\sum_j p_j^q \varepsilon_j}{\sum_j p_j^q}. \qquad (2.137)$$

Next, we partially differentiate L with respect to p_i and set it equal to zero, that is,

$$\frac{\partial L}{\partial p_i} = \frac{\partial}{\partial p_i}(S_q - \alpha \sum_j p_j - \beta U_q^{(3)}) = 0\,; \qquad (2.138)$$

now, we will place the expressions for S_q and $U_q^{(3)}$ inside the parentheses. Thus, we obtain this way:

$$\frac{\partial L}{\partial p_i} = \frac{\partial}{\partial p_i}(k\frac{1-\sum_j p_j^q}{q-1} - \alpha \sum_j p_j - \beta \frac{\sum_j p_j^q \varepsilon_j}{\sum_r p_r^q}) = 0\,, \qquad (2.139)$$

where for the sake of clarity we have introduced the dummy variable r in the expression for $U_q^{(3)}$. After differentiating inside the parentheses, it follows that

$$\frac{kq}{1-q}p_i^{q-1} - \alpha - \beta \frac{\left(qp_i^{q-1}\varepsilon_i\right)\sum_r p_r^q - (\sum_j p_j^q \varepsilon_j) q p_i^{q-1}}{\left[\sum_r p_r^q\right]^2} = 0\,; \quad (2.140)$$

then, we realize we can use again the definition of internal energy $U_q^{(3)} = \frac{\sum_j p_j^q \varepsilon_j}{\sum_r p_r^q}$ as well as introduce the new parameter $X_q = \sum_r p_r^q$. Consequently, Eq. (2.140) is expressed thus:

$$\frac{kq}{1-q}p_i^{q-1} - \alpha - \beta \frac{q\varepsilon_i p_i^{q-1} - q p_i^{q-1} U_q^{(3)}}{X_q} = 0\,, \quad (2.141)$$

after solving for p_i^{q-1}, we obtain this way:

$$p_i^{q-1} = \frac{\alpha}{\frac{kq}{1-q} - \frac{\beta}{X_q} q\varepsilon_i + \frac{\beta}{X_q} q U_q^{(3)}}\,; \qquad (2.142)$$

now, we need to know the value of α. Therefore, we will multiply Eq. (2.141) by p_i and, afterward, we will sum over all accessible states, considering the normalization condition $\sum_i p_i = 1$ as well as the parameter X_q. Thus,

$$\frac{kq}{1-q}X_q - \alpha - \beta q U_q^{(3)} + \beta q U_q^{(3)} = 0\,, \qquad (2.143)$$

after canceling the additive inverses, we solve for α,

$$\alpha = \frac{kq}{1-q}X_q\,, \qquad (2.144)$$

then, after replacing Eq. (2.144) into Eq. (2.142), we obtain the following expression:

$$p_i^{q-1} = \frac{\frac{kq}{1-q}X_q}{\frac{kq}{1-q} - \frac{\beta}{X_q}q\varepsilon_i + \frac{\beta}{X_q}qU_q^{(3)}} = \frac{X_q}{1 - \frac{\beta}{kX_q}(1-q)\varepsilon_i + \frac{\beta}{kX_q}(1-q)U_q^{(3)}}, \quad (2.145)$$

next, we factor out $-\frac{\beta}{kX_q}(1-q)$ in the denominator; we also raise p_i^{q-1} to the power -1:

$$p_i^{1-q} = \frac{1 - \frac{\beta}{kX_q}(1-q)(\varepsilon_i - U_q^{(3)})}{X_q}, \quad (2.146)$$

thus, we obtain the following probability mass function:

$$p_i = \left\{\frac{1 - \frac{\beta}{kX_q}(1-q)(\varepsilon_i - U_q^{(3)})}{X_q}\right\}^{\frac{1}{1-q}}. \quad (2.147)$$

This formula can be manipulated to contain the partition function. To this objective, in the last equation, we will apply the power of a quotient property,

$$p_i = \frac{\left\{1 - \frac{\beta}{kX_q}(1-q)(\varepsilon_i - U_q^{(3)})\right\}^{\frac{1}{1-q}}}{\{X_q\}^{\frac{1}{1-q}}}, \quad (2.148)$$

now, we will sum over all the accessible states and apply the normalization condition:

$$1 = \frac{1}{\{X_q\}^{\frac{1}{1-q}}}\sum_i\left\{1 - \frac{\beta}{kX_q}(1-q)(\varepsilon_i - U_q^{(3)})\right\}^{\frac{1}{1-q}}; \quad (2.149)$$

therefore, we can define the partition function Z_q thus:

$$Z_q = \{X_q\}^{\frac{1}{1-q}} = \sum_i\left\{1 - \frac{\beta}{kX_q}(1-q)(\varepsilon_i - U_q^{(3)})\right\}^{\frac{1}{1-q}}; \quad (2.150)$$

finally, the probability mass function in the third version of nonextensive statistical mechanics, Eq. (2.147), is expressed as follows:

$$p_i = \frac{\left\{1 - \frac{\beta}{kX_q}(1-q)(\varepsilon_i - U_q^{(3)})\right\}^{\frac{1}{1-q}}}{\sum_i\left\{1 - \frac{\beta}{kX_q}(1-q)(\varepsilon_i - U_q^{(3)})\right\}^{\frac{1}{1-q}}} = \frac{1}{Z_q}\left\{1 - \frac{\beta}{kX_q}(1-q)(\varepsilon_i - U_q^{(3)})\right\}^{\frac{1}{1-q}}. \quad (2.151)$$

On the other hand, there is another expression for p_i. This one uses the (inverse) energy parameter β'; then, to find this new expression, in the last equation we rearrange terms:

$$p_i = \frac{\left\{\left[1 + \frac{\beta}{kX_q}(1-q)U_q^{(3)}\right] - \frac{\beta}{kX_q}(1-q)\varepsilon_i\right\}^{\frac{1}{1-q}}}{\sum_i\left\{\left[1 + \frac{\beta}{kX_q}(1-q)U_q^{(3)}\right] - \frac{\beta}{kX_q}(1-q)\varepsilon_i\right\}^{\frac{1}{1-q}}}, \quad (2.152)$$

now, we can factor out $\left[1+\frac{\beta}{kX_q}(1-q)U_q^{(3)}\right]$ in the numerator and denominator; we can also cross cancel it. Consequently, the following expression holds:

$$p_i = \frac{\left\{1-\frac{\frac{\beta}{kX_q}}{\left[1+\frac{\beta}{kX_q}(1-q)U_q^{(3)}\right]}(1-q)\varepsilon_i\right\}^{\frac{1}{1-q}}}{\Sigma_i\left\{1-\frac{\frac{\beta}{kX_q}}{\left[1+\frac{\beta}{X_q}(1-q)U_q^{(3)}\right]}(1-q)\varepsilon_i\right\}^{\frac{1}{1-q}}}\,; \qquad (2.153)$$

if we name β' the common ratio,

$$\beta' = \frac{\frac{\beta}{kX_q}}{1+\frac{\beta}{kX_q}(1-q)U_q^{(3)}} = \frac{\beta}{kX_q+\beta(1-q)U_q^{(3)}}\,, \qquad (2.154)$$

then, we find the most known expression for the probability mass function,

$$p_i = \frac{\{1-(1-q)\beta'\varepsilon_i\}^{\frac{1}{1-q}}}{\Sigma_i\{1-(1-q)\beta'\varepsilon_i\}^{\frac{1}{1-q}}} = \frac{1}{Z_q}\{1-(1-q)\beta'\varepsilon_i\}^{\frac{1}{1-q}}\,. \qquad (2.155)$$

In summary, for Boltzmann-Gibbs-Shannon statistics and the three versions of nonextensive statistical mechanics, we have explicitly demonstrated how to carry out the respective processes of entropy maximization. We have also shown how different probability densities are obtained according to the constraints used. In this context, we must mention that if we use other entropies different from Tsallis entropy, with other constraints, we will find new probability mass functions. It is a subject of discussion in the scientific community because, in principle, it is possible to postulate any entropic form and obtain its respective probability mass function. However, Tsallis entropy is the most intensely investigated, with around 8000 scientific publications regarding the statistical mechanics originating from that entropy.

Demonstrating What the Lagrange Multipliers Are (in the BGS Entropy Maximization)

Previously, we have dealt with two Lagrange multipliers (α and β) in a canonical ensemble and, with three Lagrange multipliers (α, β, and γ) in a grandcanonical ensemble. Now, we will search a physical meaning for these multipliers, which appear in the BGS entropy maximization.

Demonstrating What the Lagrange Multiplier α Is (for a Canonical Ensemble)

Firstly, we start by remembering expression (2.89), $\exp\left(-1-\frac{\alpha}{k_{\mathrm{B}}}\right)=\frac{1}{\sum_i \exp\left(-\frac{\beta\varepsilon_i}{k_{\mathrm{B}}}\right)}=\frac{1}{Z_{\mathrm{c}}}$. Consequently, $\exp\left(-1-\frac{\alpha}{k_{\mathrm{B}}}\right)=\frac{1}{Z_{\mathrm{c}}}$; by taking logarithm on both sides, we have the following equation:

$$-1-\frac{\alpha}{k_{\mathrm{B}}}=\mathrm{Ln}1-\mathrm{Ln}Z_{\mathrm{c}}\,, \qquad (2.156)$$

knowing that $\mathrm{Ln}1=0$, we simplify the last equation and obtain

$$-1-\frac{\alpha}{k_{\mathrm{B}}}=-\,\mathrm{Ln}Z_{\mathrm{c}}\,; \qquad (2.157)$$

finally, solving for α, it follows that

$$\alpha=\,k_{\mathrm{B}}(\mathrm{Ln}Z_{\mathrm{c}}-1)\,. \qquad (2.158)$$

In summary, in the BGS entropy maximization, for a canonical ensemble, the Lagrange multiplier α is a parameter associated with the partition function of the corresponding thermodynamic system.

Demonstrating What the Lagrange Multiplier β Is (for a Canonical Ensemble)

Firstly, we will consider the following three thermodynamical formulae:

$$A=U-TS\,, \qquad (2.159a)$$

$$dA=-SdT-PdV\,, \quad (2.159b)$$

$$TdS=\,dU+PdV\,, \qquad (2.159c)$$

where A stands for the Helmholtz free energy; U, internal energy; T, temperature; S, entropy; P, pressure; and V, volume. Secondly, we will take into account a thermodynamical process at constant volume and constant temperature; consequently, from the last set of three equations we obtain as follows:

$$A=U-TS\,, \qquad (2.160a)$$

$$dA=0\,, \qquad (2.160b)$$

$$TdS=dU\,. \quad (2.160c)$$

Likewise, we will need an entropic expression to place it in the Eqs. (2.160a) and (2.160c). In looking for this expression, we begin by remembering Boltzmann-Gibbs-Shannon entropy, $S_{\mathrm{BGS}}=\,-k_{\mathrm{B}}\sum_i p_i\mathrm{Ln}p_i$. Thereupon, we will replace partially the following probability mass function: $p_i=\frac{1}{Z_{\mathrm{c}}}\mathrm{e}^{-\frac{\beta\varepsilon_i}{k_{\mathrm{B}}}}$. In other words,

$$S_{\mathrm{BGS}}=-k_{\mathrm{B}}\sum_i p_i\mathrm{Ln}(p_i)\,, \qquad (2.161a)$$

$$S_{\rm BGS} = -k_{\rm B}\sum_i p_i {\rm Ln}(\frac{1}{Z_{\rm c}}{\rm e}^{-\frac{\beta\varepsilon_i}{k_{\rm B}}}) , \quad (2.161b)$$

where $Z_{\rm c} = \sum_i {\rm e}^{-\frac{\beta\varepsilon_i}{k_{\rm B}}}$ is the canonical partition function. Now, in Eq. (2.161b), we will use the property of the logarithm of a quotient. Consequently, we obtain

$$S_{\rm BGS} = -k_{\rm B}\sum_i p_i [{\rm Ln}({\rm e}^{-\frac{\beta\varepsilon_i}{k_{\rm B}}}) - {\rm Ln}(Z_{\rm c})] , \quad (2.162)$$

after considering the logarithm function and exponential function are inverse functions, entropy is given as follows:

$$S_{\rm BGS} = -k_{\rm B}\sum_i p_i [-\frac{\beta\varepsilon_i}{k_{\rm B}} - {\rm Ln}(Z_{\rm c})] = k_{\rm B}\sum_i p_i [\frac{\beta\varepsilon_i}{k_{\rm B}} + {\rm Ln}(Z_{\rm c})] , \quad (2.163)$$

so, in the summation, we will apply the distributive property and, afterward, we will pull constants out. Thus, we obtain this expression:

$$S_{\rm BGS} = \beta\sum_i p_i\varepsilon_i + k_{\rm B}\sum_i p_i {\rm Ln}(Z_{\rm c}) = \beta\sum_i p_i\varepsilon_i + k_{\rm B}{\rm Ln}(Z_{\rm c})\sum_i p_i ; \quad (2.164)$$

now, we will use the definitions of internal energy and normalization condition, that is, $U = \sum_i p_i\varepsilon_i$ and $1 = \sum_i p_i$. Therefore, the last equation takes this way:

$$S_{\rm BGS} = \beta U + k_{\rm B}{\rm Ln}(Z_{\rm c}) . \quad (2.165)$$

Next, we consider $S_{\rm BGS} \equiv S$ and apply the total differential to Eq. (2.165). Thus, we get

$$dS = \beta dU + d[k_{\rm B}{\rm Ln}(Z_{\rm c})] ; \quad (2.166)$$

then, we replace this differential dS into Eq. (2.160c) and obtain as follows:

$$dU = TdS = T\{\beta dU + d[k_{\rm B}{\rm Ln}(Z_{\rm c})]\} , \quad (2.167)$$

or

$$dU = T\beta dU + d[Tk_{\rm B}{\rm Ln}(Z_{\rm c})] , \quad (2.168)$$

by comparing the left side with the right side, we arrive at the following conclusion:

$$T\beta = 1 \quad \text{and} \quad d[Tk_{\rm B}{\rm Ln}(Z_{\rm c})] = 0 , \quad (2.169)$$

finally,

$$\beta = \frac{1}{T} \quad \text{and} \quad d[k_{\rm B}T{\rm Ln}(Z_{\rm c})] = 0 . \quad (2.170)$$

Thus, in the BGS entropy maximization, for a canonical ensemble, we have demonstrated that the Lagrange operator β is associated with temperature, which is a physical quantity. Also, we want to stress that it is common to consider $\beta = \frac{1}{k_{\rm B}T}$; however, in that context, β is not the same Lagrange parameter.

On the other hand, it can be proved that $d[k_{\rm B}T{\rm Ln}(Z)] = 0$ implies the following formulae:

$$d(-A) = 0 \quad (2.171a)$$

$$A = -k_{\rm B}T{\rm Ln}(Z_{\rm c}) , \quad (2.171b)$$

where A stands for the Helmholtz free energy. Finally, from Eq. (2.171b), we have $-\frac{A}{k_B T} = \mathrm{Ln}(Z_c)$. Then, we can apply the exponential operator on both sides of this formula. Consequently, the canonical partition function is stated as follows:

$$Z_c = \exp\left(-\frac{A}{k_B T}\right). \qquad (2.172)$$

Demonstrating What the Lagrange Multiplier α Is (for a Grandcanonical Ensemble)

Firstly, we start by remembering two formulae: the grandcanonical partition function, $Z_{gc} = \sum_{i,n_i} \exp\left(-\frac{\beta\varepsilon_i + \gamma n_i}{k_B}\right)$, Eq. (2.106); and $\exp\left(-1-\frac{\alpha}{k_B}\right) = \frac{1}{Z_{gc}}$, Eq. (2.103). Next, we take the logarithm on both sides of Eq. (2.103) and get the following expression:

$$-1-\frac{\alpha}{k_B} = \mathrm{Ln}1 - \mathrm{Ln}(Z_{gc}), \qquad (2.173)$$

after considering $\mathrm{Ln}1 = 0$, the equation takes this way:

$$-1-\frac{\alpha}{k_B} = -\mathrm{Ln}(Z_{gc}); \qquad (2.174)$$

solving for α, it follows that

$$\alpha = k_B\left[\mathrm{Ln}(Z_{gc}) - 1\right]. \qquad (2.175)$$

In summary, in the BGS entropy maximization, for a grandcanonical ensemble, the Lagrange multiplier α is a parameter associated with the partition function of the corresponding system.

Demonstrating What the Lagrange Multipliers β and γ Are (for a Grandcanonical Ensemble)

Firstly, we will take into account these three thermodynamical formulae:

$$G = H - TS, \qquad (2.176a)$$

$$dG = dU + PdV + VdP - TdS - SdT, \qquad (2.176b)$$

$$TdS = dU + PdV - \mu dN \qquad (2.176c)$$

where G stands for the Gibbs free energy; H, enthalpy; T, temperature; S, entropy; U, internal energy; P, pressure; V, volume; and μ, chemical potential.

Secondly, we will consider a thermodynamical process at constant temperature and constant volume, that is, $dT = 0$ and $dV = 0$. Hence, Eqs. (2.176b) and (2.176c) are given as follows:

$$dG = dU + VdP - TdS, \qquad (2.177a)$$

$$TdS = dU - \mu dN. \qquad (2.177b)$$

Thirdly, we will search an expression for the differential of entropy dS, and afterward, we will replace it in Eq. (2.177b).

Therefore, we begin by considering Boltzmann-Gibbs-Shannon entropy, Eq. (2.161a), and, afterward, by replacing partially the probability mass function, Eq. (2.105), that is,

$$S_{\text{BGS}} = -k_{\text{B}} \sum_{i,n_i} p_i \text{Ln}(p_i) = -k_{\text{B}} \sum_{i,n_i} p_i \text{Ln}\left[\frac{\exp\left(-\frac{\beta\varepsilon_i + \gamma n_i}{k_{\text{B}}}\right)}{Z_{\text{gc}}}\right], \quad (2.178)$$

with $Z_{\text{gc}} = \sum_{i,n_i} \exp\left(-\frac{\beta\varepsilon_i + \gamma n_i}{k_{\text{B}}}\right)$. Now, we will utilize the property of the logarithm of a quotient. Thus, we obtain as follows:

$$S_{\text{BGS}} = -k_{\text{B}} \sum_{i,n_i} p_i [\text{Ln}\left\{\exp\left(-\frac{\beta\varepsilon_i + \gamma n_i}{k_{\text{B}}}\right)\right\} - \text{Ln}(Z_{\text{gc}})], \quad (2.179)$$

after considering the exponential function and the logarithm function are inverse functions, it follows that

$$S_{\text{BGS}} = -k_{\text{B}} \sum_{i,n_i} p_i [-\frac{\beta\varepsilon_i + \gamma n_i}{k_{\text{B}}} - \text{Ln}(Z_{\text{gc}})], \quad (2.180)$$

by multiplying we obtain this:

$$S_{\text{BGS}} = \beta \sum_{i,n_i} p_i \varepsilon_i + \gamma \sum_{i,n_i} p_i n_i + k_{\text{B}} \sum_{i,n_i} p_i \text{Ln}(Z_{\text{gc}}); \quad (2.181)$$

next, we will sequentially use the definitions of internal energy, number of particles, and normalization condition; that is, $U = \sum_i p_i \varepsilon_i$, $N = \sum_i p_i n_i$, and $1 = \sum_i p_i$. Therefore, the last equation is reduced to the following one:

$$S_{\text{BGS}} = \beta U + \gamma N + k_{\text{B}} \text{Ln}(Z_{\text{gc}}); \quad (2.182)$$

now, we will consider $S_{\text{BGS}} \equiv S$ and take the total differential. Thus, we obtain as follows:

$$dS = \beta dU + \gamma dN + d[k_{\text{B}} \text{Ln}(Z_{\text{gc}})]; \quad (2.183)$$

then, we can replace this equation into Eq. (2.177b), which becomes this way:

$$TdS = T\{\beta dU + \gamma dN + d[k_{\text{B}} \text{Ln}(Z_{\text{gc}})]\} = dU - \mu dN; \quad (2.184)$$

next, in the middle part above, we will apply the distributive property. Hence,

$$T\beta dU + T\gamma dN + d[Tk_{\text{B}} \text{Ln}(Z_{\text{gc}}) = dU - \mu dN, \quad (2.185)$$

by comparing terms on the left side and right side, respectively, we arrive at the following conclusions:

$$T\beta = 1, \quad T\gamma = -\mu, \quad \text{and} \quad d[Tk_{\text{B}} \text{Ln}(Z_{\text{gc}})] = 0, \quad (2.186)$$

or

$$\beta = \frac{1}{T}, \quad \gamma = -\frac{\mu}{T} \text{ and } \quad d[k_{\text{B}} T \text{Ln}(Z_{\text{gc}})] = 0. \quad (2.187)$$

Consequently, in the BGS entropy maximization, for a grandcanonical ensemble, we have again demonstrated that the Lagrange operator β is associated with temperature; likewise, we have discovered that the

Lagrange parameter γ is associated with temperature and chemical potential, too; γ has the same dimension of k_{B}.

On the other hand, it can be proved that the result $d\left[k_{\mathrm{B}}T\mathrm{Ln}\left(Z_{\mathrm{gc}}\right)\right]=0$ implies the following conclusion:

$$\Phi=-k_{\mathrm{B}}T\mathrm{Ln}\left(Z_{\mathrm{gc}}\right), \qquad (2.188)$$

where Φ is the grand potential, which is a thermodynamical potential. The last formula is similar to Eq. (2.171b), which is valid for a canonical ensemble. We must also bear in mind that, in classical thermodynamics, the grand potential is defined in this manner:

$$\Phi_{T,\mu}=E-TS-\mu N\,. \qquad (2.189)$$

Finally, from Eq. (2.188), we can obtain $-\frac{1}{k_{\mathrm{B}}T}\Phi=\mathrm{Ln}\left(Z_{\mathrm{gc}}\right)$ and apply the exponential on both sides of this expression. Thus, the grandcanonical partition function is given by

$$Z_{\mathrm{gc}}=\exp\left(-\frac{1}{k_{\mathrm{B}}T}\Phi\right); \qquad (2.190)$$

this formula is analogous to Eq. (2.172), which is valid for a canonical ensemble.

Demonstrating How to Obtain Boltzmann Entropy from Boltzmann-Gibbs-Shannon Entropy

Firstly, Boltzmann entropy is formulated thus:

$$S_{\mathrm{Boltz.}}=k_{\mathrm{B}}\mathrm{Ln}\Omega\,, \qquad (2.191)$$

where Ω is the number of accessible states for the system, and k_{B} is the Boltzmann constant. Also, we know Boltzmann-Gibbs-Shannon entropy has the following expression:

$$S_{\mathrm{BGS}}=-k_{\mathrm{B}}\sum_i[p_i\mathrm{Ln}p_i]\,. \qquad (2.192)$$

Therefore, we begin the demonstration by remembering about the probability mass function $p_i=1/\Omega$ (for a microcanonical ensemble). Next, we replace this formula into Eq. (2.192), which is expressed this way:

$$S_{\mathrm{BGS}}=-k_{\mathrm{B}}\sum_i\left[\frac{1}{\Omega}\mathrm{Ln}\left(\frac{1}{\Omega}\right)\right]; \qquad (2.193)$$

by using the rule for the logarithm of a quotient, this entropy is stated as follows:

$$S_{\mathrm{BGS}}=k_{\mathrm{B}}\sum_i\left[\frac{1}{\Omega}\mathrm{Ln}(\Omega)\right]; \qquad (2.194)$$

now, we use the properties of summation and obtain the following expression:

$$S_{\text{BGS}} = k_{\text{B}} \frac{\text{Ln}(\Omega)}{\Omega} \sum_i 1 \,, \qquad (2.195)$$

then, we sum constant 1 over all the accessible states. Thereupon, we obtain this way:

$$S_{\text{BGS}} = k_{\text{B}} \frac{\text{Ln}(\Omega)}{\Omega} \Omega \,, \qquad (2.196)$$

after cross canceling Ω, we finally get

$$S_{\text{BGS}} = k_{\text{B}} \text{Ln}(\Omega) = S_{\text{Boltz.}} \,. \qquad (2.197)$$

Demonstrations of Several Properties of the *q*-Exponential Function

Finding Out a Formula for the q-Exponential Function of the Sum of Two Functions

We will search a formula for $\exp_q(y_1 + y_2)$. Then, we start by considering the definition of the corresponding q-exponential function:

$$\exp_q(y_1 + y_2) = [1 + (1 - q)(y_1 + y_2)]^{\frac{1}{1-q}} \,, \qquad (2.198)$$

after multiplying the binomials $(1 - q)$ and $(y_1 + y_2)$,

$$\exp_q(y_1 + y_2) = [1 + (1 - q)y_1 + (1 - q)y_2]^{\frac{1}{1-q}} \,; \qquad (2.199)$$

now, we add and subtract 1 inside the brackets, and after rearranging terms, we obtain

$$\exp_q(y_1 + y_2) = [1 + (1 - q)y_1 + 1 + (1 - q)y_2 - 1]^{\frac{1}{1-q}} \,, \qquad (2.200)$$

by using again the definition of the q-exponential function, we get this result:

$$\exp_q(y_1 + y_2) = \left[\exp_q(y_1)^{1-q} + \exp_q(y_2)^{1-q} - 1\right]^{\frac{1}{1-q}} \,. \qquad (2.201)$$

If we want to maintain similarities with the properties of standard mathematics, we realize we must introduce a new multiplication operation this way:

$$\exp_q(y_1 + y_2) = \exp_q(y_1) \otimes_q \exp_q(y_2) \,, \qquad (2.202)$$

where obviously

$$\exp_q(y_1) \otimes_q \exp_q(y_2) = \left[\exp_q(y_1)^{1-q} + \exp_q(y_2)^{1-q} - 1\right]^{\frac{1}{1-q}} \,; \qquad (2.203)$$

in summary, this new multiplication is named the q-product and is defined thus:

$$a \otimes_q b = \begin{cases} [a^{1-q} + b^{1-q} - 1]^{\frac{1}{1-q}}, & \text{if } a > 0,\ b > 0 \text{ and } a^{1-q} + b^{1-q} - 1 > 0 \\ 0, & \text{elsewhere} \end{cases} \,, \qquad (2.204)$$

On the other hand, for $q = 1$, we have a limit,

$$a \otimes_1 b = \lim_{q\to 1}[a^{1-q} + b^{1-q} - 1]^{\frac{1}{1-q}}, \qquad (2.205)$$

adding and subtracting 1 inside the brackets, it follows that

$$a \otimes_1 b = \lim_{q\to 1}[1 + a^{1-q} - 1 + b^{1-q} - 1]^{\frac{1}{1-q}}, \qquad (2.206)$$

now, we will multiply $a^{1-q} - 1$ and $b^{1-q} - 1$ by $(1-q)/(1-q)$, respectively. Thus,

$$a \otimes_1 b = \lim_{q\to 1}\left[1 + (1-q)\frac{a^{1-q}-1}{1-q} + (1-q)\frac{b^{1-q}-1}{1-q}\right]^{\frac{1}{1-q}}, (2.207)$$

after factoring out $(1-q)$, we have

$$a \otimes_1 b = \lim_{q\to 1}\left[1 + (1-q)\left(\frac{a^{1-q}-1}{1-q} + \frac{b^{1-q}-1}{1-q}\right)\right]^{\frac{1}{1-q}}, \quad (2.208)$$

and by applying the definition of the q-logarithm function, we get

$$a \otimes_1 b = \lim_{q\to 1}\left[1 + (1-q)\left(\mathrm{Ln}_q a + \mathrm{Ln}_q b\right)\right]^{\frac{1}{1-q}}; \qquad (2.209)$$

then, we use again the definition of the q-exponential function. Thus, we obtain

$$a \otimes_1 b = \lim_{q\to 1}\left[\exp_q\left(\mathrm{Ln}_q a + \mathrm{Ln}_q b\right)\right], \qquad (2.210)$$

after evaluating the limit,

$$a \otimes_1 b = \exp_1(\mathrm{Ln}_1 a + \mathrm{Ln}_1 b) = \exp(\mathrm{Ln} a + \mathrm{Ln} b); \qquad (2.211)$$

now, according to the property of the addition of two logarithms, it becomes thus:

$$a \otimes_1 b = \exp(\mathrm{Ln}(ab)) = ab, \qquad (2.212)$$

where we have considered that the exponential and the logarithm are inverse functions. Thus,

$$a \otimes_1 b = a \cdot b, \qquad (2.213)$$

that is, for $q = 1$, the q-product is the standard multiplication operation,

$$\otimes_1 \equiv \cdot \qquad (2.214)$$

Consequently, for $q = 1$, we also have

$$\begin{aligned}\exp_1(y_1 + y_2) &= \exp_1(y_1) \otimes_1 \exp_1(y_2)\\ &= \exp(y_1)\exp(y_2), \qquad (2.215)\end{aligned}$$

which is the standard result.

Finding Out a Formula for the q-Exponential Function of the Sum of Three Functions

We want to find a formula for $\exp_q(y_1 + y_2 + y_3)$, wherein we have three variables. Then, we have the following expression:

$$\exp_q(y_1 + y_2 + y_3) = [1 + (1-q)(y_1 + y_2 + y_3)]^{\frac{1}{1-q}}, (2.216)$$

multiplying the second addend on the right side, we have

$$\exp_q(y_1+y_2+y_3)=[1+(1-q)y_1+(1-q)y_2+(1-q)y_3]^{\frac{1}{1-q}}, \quad (2.217)$$

adding and subtracting 2, and grouping terms, we get

$$\exp_q(y_1+y_2+y_3)=[1+(1-q)y_1+1+(1-q)y_2+1+(1-q)y_3-2]^{\frac{1}{1-q}}, \quad (2.218)$$

using the definition of the q-exponential, it follows that

$$\exp_q(y_1+y_2+y_3)=\left[\exp_q(y_1)^{1-q}+\exp_q(y_2)^{1-q}+\exp_q(y_3)^{1-q}-2\right]^{\frac{1}{1-q}}. \quad (2.219)$$

Therefore, similarly to the case of two variables, we define

$$\exp_q(y_1+y_2+y_3)=\exp_q(y_1)\otimes_q\exp_q(y_2)\otimes_q\exp_q(y_3), \quad (2.220)$$

with

$$\exp_q(y_1)\otimes_q\exp_q(y_2)\otimes_q\exp_q(y_3)=$$

$$=\left[\exp_q(y_1)^{1-q}+\exp_q(y_2)^{1-q}+\exp_q(y_3)^{1-q}-2\right]^{\frac{1}{1-q}}. \quad (2.221)$$

On the other hand, for $q=1$, we have

$$\exp_1(y_1+y_2+y_3)=\exp_1(y_1)\otimes_1\exp_1(y_2)\otimes_1\exp_1(y_3)$$
$$=\exp(y_1).\exp(y_2).\exp(y_3) \quad (2.222)$$

or

$$\exp(y_1+y_2+y_3)=\exp(y_1)\exp(y_2)\exp(y_3), \quad (2.223)$$

which is the standard result.

Demonstrating a Formula for the q-Exponential of the Sum of n Functions

Based on the previous results, it can be proposed that

$$\exp_q(y_1+y_2+\cdots+y_n)=\exp_q(y_1)\otimes_q\exp_q(y_2)\otimes_q\ldots\otimes_q\exp_q(y_n) \quad (2.224)$$

and

$$\exp_q(y_1)\otimes_q\exp_q(y_2)\otimes_q\ldots\otimes_q\exp_q(y_n)=$$

$$=\left[\exp_q(y_1)^{1-q}+\exp_q(y_2)^{1-q}+\cdots+\exp_q(y_n)^{1-q}-(n-1)\right]^{\frac{1}{1-q}}, \quad (2.225)$$

where $n\geq 2$ and $n\in\mathbb{N}$. To demonstrate it, we will use the method of mathematical proof called mathematical induction.

Thus, for $n=2$, we have the following expression:

$$\exp_q(y_1+y_2)=\exp_q(y_1)\otimes_q\exp_q(y_2). \quad (2.226)$$

but this formula has been already proved previously. Additionally, we already know that

$$\exp_q(y_1)\otimes_q\exp_q(y_2)=\left[\exp_q(y_1)^{1-q}+\exp_q(y_2)^{1-q}-1\right]^{\frac{1}{1-q}}, \quad (2.227)$$

which can be expressed this way:

$$\exp_q(y_1)\otimes_q\exp_q(y_2)=\left[\exp_q(y_1)^{1-q}+\exp_q(y_2)^{1-q}-(2-1)\right]^{\frac{1}{1-q}}; \quad (2.228)$$

therefore, for $n=2$, it has already been proved.

For $n = h$, the formula is given as follows:
$$\exp_q(y_1 + y_2 + \cdots + y_h) = \exp_q(y_1) \otimes_q \ldots \otimes_q \exp_q(y_h)\,, \qquad (2.229)$$
where
$$\exp_q(y_1) \otimes_q \ldots \otimes_q \exp_q(y_h) = $$
$$= \left[\exp_q(y_1)^{1-q} + \cdots + \exp_q(y_h)^{1-q} - (h-1)\right]^{\frac{1}{1-q}}, \qquad (2.230)$$
following the principle of mathematical induction, we will assume that both formulae are true.

For $n = h + 1$, we have
$$\exp_q(y_1 + \cdots + y_{h+1}) = \exp_q(y_1) \otimes_q \ldots \otimes_q \exp_q(y_{h+1})\,. \quad (2.231a)$$
$$\exp_q(y_1) \otimes_q \ldots \otimes_q \exp_q(y_{h+1}) =$$
$$= \left[\exp_q(y_1)^{1-q} + \cdots + \exp_q(y_{h+1})^{1-q} - ((h+1)-1)\right]^{\frac{1}{1-q}}. \quad (2.231b)$$
We begin by demonstrating Eq. (2.231a). Then, on the left side of it, we apply the definition of the q-exponential,
$$\exp_q(y_1 + \cdots + y_h + y_{h+1}) = [1 + (1-q)(y_1 + \cdots + y_h + y_{h+1})]^{\frac{1}{1-q}}, \quad (2.232)$$
inside the brackets, we multiply $(1-q)(y_1 + \cdots + y_h + y_{h+1})$, but previously we define the auxiliary parameter $D_0 = \exp_q(y_1 + \cdots + y_h + y_{h+1})$; consequently,
$$D_0 = [1 + (1-q)(y_1 + \cdots + y_h) + (1-q)(y_{h+1})]^{\frac{1}{1-q}}, \qquad (2.233)$$
by adding and subtracting 1 inside the brackets, we have $D_0 = [1 + (1-q)(y_1 + \cdots + y_h) + 1 + (1-q)(y_{h+1}) - 1]^{\frac{1}{1-q}}$. After using the definition of the q-exponential, we obtain
$$D_0 = \left[\exp_q(y_1 + \cdots + y_h)^{1-q} + \exp_q(y_{h+1})^{1-q} - 1\right]^{\frac{1}{1-q}}, \qquad (2.234)$$
naming $a = y_1 + \cdots + y_h$ and $b = y_{h+1}$, we obtain
$$D_0 = \left[\exp_q(a)^{1-q} + \exp_q(b)^{1-q} - 1\right]^{\frac{1}{1-q}}. \qquad (2.235)$$
Thereupon, it has already been defined that
$$\left[\exp_q(a)^{1-q} + \exp_q(b)^{1-q} - 1\right]^{\frac{1}{1-q}} \equiv \exp_q(a) \otimes_q \exp_q(b)\,, \qquad (2.236)$$
consequently,
$$D_0 = \exp_q(y_1 + \cdots + y_h) \otimes_q \exp_q(y_{h+1})\,. \qquad (2.237)$$
Now, we will use the inductive hypothesis, Eq. (2.229), which was given by
$$\exp_q(y_1 + \cdots + y_h) = \exp_q(y_1) \otimes_q \exp_q(y_2) \otimes_q \ldots \otimes_q \exp_q(y_h)\,; \quad (2.238)$$
thus, replacing this last equation into the penultimate equation, we obtain
$$D_0 = \exp_q(y_1) \otimes_q \exp_q(y_2) \otimes_q \ldots \otimes_q \exp_q(y_h) \otimes_q \exp_q(y_{h+1}). \quad (2.239)$$
Then, we have proved Eq. (2.231a) and, consequently, Eq. (2.224).

Therefore, it only lacks to prove Eq. (2.231b); rewriting it,
$\exp_q(y_1) \otimes_q \dots \otimes_q \exp_q(y_{h+1}) =$

$$\left[\exp_q(y_1)^{1-q} + \dots + \exp_q(y_{h+1})^{1-q} - ((h+1)-1)\right]^{\frac{1}{1-q}}, \quad (2.240)$$

then, on the left side, we introduce the following auxiliary parameter: $D_1 = \exp_q(y_1) \otimes_q \dots \otimes_q \exp_q(y_{h+1})$; we also group terms and obtain
$D_1 = \left[\exp_q(y_1) \otimes_q \dots \otimes_q \exp_q(y_h)\right] \otimes_q \exp_q(y_{h+1}) =$

$$= c \otimes_q d = [c^{1-q} + d^{1-q} - 1)]^{\frac{1}{1-q}}, \quad (2.241)$$

where we have introduced

$$c = \exp_q(y_1) \otimes_q \dots \otimes_q \exp_q(y_h) \quad \text{and} \quad d = \exp_q(y_{h+1})\,; \quad (2.242)$$

then, by using the hypothesis of Eq. (2.230), we have this way:

$$D_1 = \left[\left\{\left[\exp_q(y_1)^{1-q} + \dots + \exp_q(y_h)^{1-q} - (h-1)\right]^{\frac{1}{1-q}}\right\}^{1-q} + \exp_q(y_{h+1})^{1-q} - 1\right]^{\frac{1}{1-q}}, \quad (2.243)$$

simplifying,

$$D_1 = \left[\exp_q(y_1)^{1-q} + \dots + \exp_q(y_h)^{1-q} - (h-1) + \exp_q(y_{h+1})^{1-q} - 1)\right]^{\frac{1}{1-q}}, \quad (2.244)$$

and after rearranging on the right side,

$$D_1 = \left[\exp_q(y_1)^{1-q} + \dots + \exp_q(y_h)^{1-q} + \exp_q(y_{h+1})^{1-q} - (h-1) - 1)\right]^{\frac{1}{1-q}}, \quad (2.245)$$

after considering that $-(h-1) - 1 = -[(h+1) - 1]$, it follows that

$$D_1 = \left\{\exp_q(y_1)^{1-q} + \dots + \exp_q(y_h)^{1-q} + \exp_q(y_{h+1})^{1-q} - [(h+1) - 1]\right\}^{\frac{1}{1-q}}, \quad (2.246)$$

that is, we have demonstrated Eq. (2.231b) and, consequently, Eq. (2.225)

Finding Out a Formula for the Product of Two q-Exponentials

We search a formula for the following product: $\exp_q(y_1)\exp_q(y_2)$. Then, in accordance with the definition of the q-exponential, we have thus:

$$\exp_q(y_1)\exp_q(y_2) = [1 + (1 - q)(y_1)]^{\frac{1}{1-q}}[1 + (1 - q)(y_2)]^{\frac{1}{1-q}}\,; \quad (2.247)$$

now, on the right side, we will use the power of a product property. Therefore,

$$\exp_q(y_1)\exp_q(y_2) = \{[1 + (1 - q)y_1][1 + (1 - q)y_2]\}^{\frac{1}{1-q}}, \quad (2.248)$$

after multiplying inside the braces, we have

$$\exp_q(y_1)\exp_q(y_2) = \{1 + (1 - q)y_2 + (1 - q)y_1 + (1 - q)y_1(1 - q)y_2\}^{\frac{1}{1-q}}, \quad (2.249)$$

rearranging terms on the right side,

$$\exp_q(y_1)\exp_q(y_2) = \{1 + (1 - q)y_1 + (1 - q)y_2 + (1 - q)^2 y_1 y_2\}^{\frac{1}{1-q}}, \quad (2.250)$$

by factoring out $(1 - q)$ of the second and the third addend, we obtain

$$\exp_q(y_1)\exp_q(y_2) = \{1 + (1 - q)(y_1 + y_2) + (1 - q)^2 y_1 y_2\}^{\frac{1}{1-q}}\,; \quad (2.251)$$

now, we again factor out $(1 - q)$,

$$\exp_q(y_1)\exp_q(y_2) = \{1 + (1 - q)[y_1 + y_2 + (1 - q)y_1 y_2]\}^{\frac{1}{1-q}}, \quad (2.252)$$

we recognize the right side of the last equation as a q-exponential, that is,

$$\exp_q(y_1)\exp_q(y_2) = \exp_q[y_1 + y_2 + (1 - q)y_1y_2], \quad (2.253)$$

Finally, in order to maintain similarities with the properties of standard mathematics, we must introduce a new operation:

$$\exp_q(y_1)\exp_q(y_2) = \exp_q(y_1 \oplus_q y_2), \quad (2.254)$$

this new way of adding is named the q-sum and is defined thus:

$$y_1 \oplus_q y_2 = y_1 + y_2 + (1 - q)y_1y_2, \quad (2.255)$$

or more generally, for any functions a and b:

$$a \oplus_q b = a + b + (1 - q)ab. \quad (2.256)$$

On the other hand, for $q = 1$, we have

$$a \oplus_1 b = a + b + (1 - 1)ab, \quad (2.257)$$

which implies that

$$a \oplus_1 b = a + b \quad \text{and} \quad \oplus_1 \equiv +, \quad (2.258)$$

that is, for $q = 1$, we recover the usual addition operation

Finding Out a Formula for the Product of Three q-Exponentials

We start by applying the last result; then,

$$\exp_q(y_1)\exp_q(y_2)\exp_q(y_3) = \exp_q[y_1 + y_2 + (1 - q)y_1y_2]\exp_q(y_3), \quad (2.259)$$

and according to the definition of the q-exponential function,

$$\exp_q(y_1)\exp_q(y_2)\exp_q(y_3) = \{1 + (1 - q)[y_1 + y_2 + (1 - q)y_1y_2]\}^{\frac{1}{1-q}}[1 + (1 - q)y_3]^{\frac{1}{1-q}} \quad (2.260)$$

now, on the right side, we will use the product of powers property along with the corresponding internal multiplications. However, to simplify the notation, we will previously introduce the following auxiliary parameter $D_2 = \exp_q(y_1)\exp_q(y_2)\exp_q(y_3)$. Therefore,

$$D_2 = \{1 + (1 - q)y_3 + (1 - q)[y_1 + y_2 + (1 - q)y_1y_2]\,[1 + (1 - q)y_3]\}^{\frac{1}{1-q}}, \quad (2.261)$$

and by multiplying terms again, it follows that

$$D_2 = \{1 + (1 - q)y_3 + (1 - q)[y_1 + y_2 + (1 - q)y_1y_2] + (1 - q)[y_1 + y_2 + (1 - q)y_1y_2](1 - q)y_3\}^{\frac{1}{1-q}} \quad (2.262)$$

multiplying once more,

$$D_2 = \{1 + (1 - q)y_3 + (1 - q)(y_1 + y_2) + (1 - q)^2y_1y_2 + (1 - q)^2[y_1y_3 + y_2y_3 + (1 - q)y_1y_2y_3]\}^{\frac{1}{1-q}} \quad (2.263)$$

by factoring out $(1 - q)$ and $(1 - q)^2$, we have

$$D_2 = \{1 + (1 - q)(y_1 + y_2 + y_3) + (1 - q)^2(y_1y_2 + y_1y_3 + y_2y_3) + (1 - q)^3y_1y_2y_3\}^{\frac{1}{1-q}} \quad (2.264)$$

by factoring out $(1 - q)$ again, it becomes this way:

$$D_2 = \{1 + (1 - q)[y_1 + y_2 + y_3 + (1 - q)(y_1y_2 + y_1y_3 + y_2y_3) + (1 - q)^2y_1y_2y_3]\}^{\frac{1}{1-q}}; \quad (2.265)$$

next, we can identify the q-exponential function on the right side:

$$D_2 = \exp_q[y_1 + y_2 + y_3 + (1 - q)(y_1y_2 + y_1y_3 + y_2y_3) + (1 - q)^2y_1y_2y_3]. \quad (2.266)$$

Likewise, we also identify the q-sum of y_1, y_2, and y_3 inside the brackets: $D_2 = \exp_q(y_1 \oplus_q y_2 \oplus_q y_3)$. Subsequently, we conclude that $\exp_q(y_1)\exp_q(y_2)\exp_q(y_3) = \exp_q(y_1 \oplus_q y_2 \oplus_q y_3)$ is the formula we searched for.

On the other hand, for $q = 1$, we formulate thus:

$$\exp_1(y_1)\exp_1(y_2)\exp_1(y_3) = \exp_1(y_1 \oplus_1 y_2 \oplus_1 y_3), \quad (2.267)$$

after simplifying, we obtain this way:

$$\exp(y_1)\exp(y_2)\exp(y_3) = \exp[y_1 + y_2 + y_3], \quad (2.268)$$

which is the established result.

Finding Out a Formula for the q-Exponential of the Difference of Two Functions

For this purpose, we will use the definition of the q-exponential function:

$$\exp_q(y_1 - y_2) = [1 + (1-q)(y_1 - y_2)]^{\frac{1}{1-q}} = [1 + (1-q)y_1 - (1-q)y_2]^{\frac{1}{1-q}}, \quad (2.269)$$

then, we will express $(1-q)y_1$ and $(1-q)y_2$ as a function of $\exp_q(y_1)$ and $\exp_q(y_2)$, respectively. Thus, we have as follows:

$$\exp_q(y_1 - y_2) = \left\{\exp_q(y_1)^{1-q} - \left[\exp_q(y_2)^{1-q} - 1\right]\right\}^{\frac{1}{1-q}}$$

$$= \left[\exp_q(y_1)^{1-q} - \exp_q(y_2)^{1-q} + 1\right]^{\frac{1}{1-q}}, \quad (2.270)$$

now, we can define a new operation called the q-division

$$\exp_q(y_1 - y_2) = \exp_q(y_1) \oslash_q \exp_q(y_2), \quad (2.271)$$

where obviously it follows that

$$\exp_q(y_1) \oslash_q \exp_q(y_2) = \left[\exp_q(y_1)^{1-q} - \exp_q(y_2)^{1-q} + 1\right]^{\frac{1}{1-q}}; \quad (2.272)$$

more generally, we can write out

$$a \oslash_q b = [a^{1-q} - b^{1-q} + 1]^{\frac{1}{1-q}} \quad \text{with} \quad a^{1-q} - b^{1-q} + 1 > 0. \quad (2.273)$$

Besides, for $q = 1$, Eq. (2.271) is expressed as $\exp_1(y_1 - y_2) = \exp_1(y_1) \oslash_1 \exp_1(y_2)$, that is, $\exp(y_1 - y_2) = \exp(y_1) \oslash_1 \exp(y_2)$. Therefore, we realize $\oslash_1 \equiv /$ is a necessary equivalence. Below, we will prove it.

Thus, for the q-division with $q = 1$, we have

$$a \oslash_1 b = \mathbf{Lim}_{q \to 1}[a^{1-q} - b^{1-q} + 1]^{\frac{1}{1-q}}, \quad (2.274)$$

adding and subtracting 1 inside the brackets,

$$a \oslash_1 b = \mathbf{Lim}_{q \to 1}[1 - 1 + a^{1-q} - b^{1-q} + 1]^{\frac{1}{1-q}} = \mathbf{Lim}_{q \to 1}[1 + (a^{1-q} - 1) - (b^{1-q} - 1)]^{\frac{1}{1-q}}, \quad (2.275)$$

after multiplying and dividing by $(1-q)$ the second and third addends, we have

$$a \oslash_1 b = \mathbf{Lim}_{q \to 1}\left[1 + (1-q)\frac{a^{1-q} - 1}{1-q} - (1-q)\left(\frac{b^{1-q} - 1}{1-q}\right)\right]^{\frac{1}{1-q}}, \quad (2.276)$$

factoring out $(1-q)$,

$$a \oslash_1 b = \mathbf{Lim}_{q \to 1}\left[1 + (1-q)\left(\frac{a^{1-q} - 1}{1-q} - \frac{b^{1-q} - 1}{1-q}\right)\right]^{\frac{1}{1-q}} = \mathbf{Lim}_{q \to 1}\left[1 + (1-q)\left(\mathrm{Ln}_q a - \mathrm{Ln}_q b\right)\right]^{\frac{1}{1-q}} \quad (2.277)$$

and after applying the definition of the q-exponential inside the last brackets, we obtain

$$a \oslash_1 b = \underset{q\to 1}{\mathbf{Lim}}\left[\exp_q(\mathrm{Ln}_q a - \mathrm{Ln}_q b)\right] = \exp(\mathrm{Ln}a - \mathrm{Ln}b)\ , \quad (2.278)$$

now, we will apply the property of the difference of two logarithms. Consequently,

$$a \oslash_1 b = \exp\left(\mathrm{Ln}\left(\frac{a}{b}\right)\right), \qquad (2.279)$$

which implies that

$$a \oslash_1 b = a/b \ \text{ and } \ \oslash_1 \equiv / \ , \quad (2.280)$$

that is, for $q = 1$, we have recovered the classical division operation.

Finding Out a Formula for the Division of two q-Exponentials

we will start by using the definition of the q-exponential function. Thus,

$$\frac{\exp_q(y_1)}{\exp_q(y_2)} = \frac{[1+(1-q)y_1]^{\frac{1}{1-q}}}{[1+(1-q)y_2]^{\frac{1}{1-q}}} = \left[\frac{1+(1-q)y_1}{1+(1-q)y_2}\right]^{\frac{1}{1-q}}, \quad (2.281)$$

adding and subtracting $(1-q)y_2$ in the numerator, we have as follows:

$$\frac{\exp_q(y_1)}{\exp_q(y_2)} = \left[\frac{1+(1-q)y_1+(1-q)y_2-(1-q)y_2}{1+(1-q)y_2}\right]^{\frac{1}{1-q}} = \left[1 + \frac{(1-q)y_1-(1-q)y_2}{1+(1-q)y_2}\right]^{\frac{1}{1-q}}, \ (2.282)$$

by factoring out $(1-q)$ in the numerator of the second addend (the ratio), it follows that

$$\frac{\exp_q(y_1)}{\exp_q(y_2)} = \left[1 + (1-q)\frac{y_1-y_2}{1+(1-q)y_2}\right]^{\frac{1}{1-q}}; \qquad (2.283)$$

next, we realize we have arrived at other q-exponential, namely,

$$\frac{\exp_q(y_1)}{\exp_q(y_2)} = \exp_q\left[\frac{y_1-y_2}{1+(1-q)y_2}\right]. \qquad (2.284)$$

To maintain similarity with the standard formulae, we can introduce a new operation called the q-difference, that is,

$$y_1 \ominus_q y_2 \equiv \frac{y_1-y_2}{1+(1-q)y_2}; \qquad (2.285)$$

therefore, we obtain this way:

$$\frac{\exp_q(y_1)}{\exp_q(y_2)} = \exp_q(y_1 \ominus_q y_2); \quad (2.286)$$

more generally, we can express the q-difference this way:

$$a \ominus_q b = \frac{a-b}{1+(1-q)b}; \qquad (2.287)$$

besides, for example, for $q = 1$, we obtain this way:

$$a \ominus_1 b = \frac{a-b}{1+(1-1)b} = \frac{a-b}{1}, \ (2.288)$$

which implies

$$a \ominus_1 b = a - b \ \text{ and } \ \ominus_1 \equiv - \qquad (2.289)$$

that is, for $q = 1$, we recover the usual subtraction operation.

In this chapter, the basic concepts of nonextensive statistical mechanics have been outlined. In the context of this theory, in Chapter Three, a new method for consistently calculating the mean values of a system $A + B$ will be provided. Thereupon, the new method is concordant with the concept of partial trace.

CHAPTER THREE

AN ALTERNATIVE METHOD TO CALCULATE MEAN VALUES OF COMPOUND SYSTEMS

The calculations of magnetization in manganites require computing the one of a correlated system A + B because, in these materials, manganese presents with mixed valence (Mn^{3+} and Mn^{4+}). These ions have spins S_A = 2 and S_B = 3/2, respectively (remember that S stands for spin and *S*, entropy!). Thus, we have two interacting magnetic sublattices. When developing this topic, a problem arises in the calculations: How to correctly calculate the mean values of these subsystems. Even though we might initially consider the previous spins, the analytical calculations for a correlated two spin-½ system will be shown since these calculations are very didactic and illustrative. Specifically, the quantum states of the just mentioned system can be represented in the 4x4 full Hilbert space, whereas the Mn^{3+} and Mn^{4+} states require the 20x20 full Hilbert space.

Consequently, to completely understand the above problem as well as its solution, we will proceed as follows: firstly, we will show general formulae to calculate quantum mean values in the Hilbert spaces; secondly, we will analytically resolve the aforementioned two spin-½ system. Likewise, the problem and its solution will be shown through computer calculations, some of which were performed by using the manganese spins S_A = 2 and S_B = 3/2. Furthermore, we will also introduce a section on the escort distributions and compound systems.

A Formal Derivation from the Full Hilbert Space Toward the Hilbert Subspaces

We can use the partial trace to go from mean values formulae in the full Hilbert space to mean values formulae in the Hilbert subspaces. The derivation below is a simplification of a rigorous procedure, which is analogous to the one found in the famous scholarly book *Quantum mechanics* by Claude Cohen-Tannoudji, Bernard Diu, and Frank Laloë (evidently, the procedure done by them is only valid for $q = 1$). Besides, the derivation below is independent of the physical system, the definition of temperature, and the values of q, too. Even more, the expression of the density matrix is not relevant. Thereupon, the derivation below can be applied to any density matrix of any statistical mechanics that defines the mean values in the same way as the third version of nonextensive statistical mechanics does.

Thus, in Figure 3.1, we realize the derivation consists of four steps to go from the formulae in the full Hilbert space to the formulae in the Hilbert subspaces. In this graphic, E_{A} and E_{B} are the matrices defined by $E_{\mathrm{A}} = \mathbf{Tr}_{\mathrm{B}}(\rho^q)$ and $E_{\mathrm{B}} = \mathbf{Tr}_{\mathrm{A}}(\rho^q)$; O_{A} and O_{B} are the respective matrix representations of the operators $\hat{O}_{\mathrm{A}}$ and $\hat{O}_{\mathrm{B}}$. Likewise, $\mathcal{O}_{\mathrm{A}}$ and $\mathcal{O}_{\mathrm{B}}$ are the matrices for the same observables in the Hilbert subspaces; in fact, $O_{\mathrm{A}} = \mathcal{O}_{\mathrm{A}} \otimes \mathbb{1}_{\mathrm{B}}$ and $O_{\mathrm{B}} = \mathbb{1}_{\mathrm{A}} \otimes \mathcal{O}_{\mathrm{B}}$, where $\mathbb{1}_A$ and $\mathbb{1}_B$ are the identity matrices in subspaces A and B, respectively.

Step **I** represents the mean value of an observable of the subsystem A (or B) in the full Hilbert space. Step **II** is obtained by taking the corresponding partial trace. Step **III** is achieved from step **II**, whenever we assume the q-statistical independence $\rho^q = \rho_{\mathrm{A}}^q \otimes \rho_{\mathrm{B}}^q$. Step **IV** contains the simplified formulae in the Hilbert subspaces.

I		**II**		**III**		**IV**
$\dfrac{\mathbf{Tr}_{(\mathrm{A,B})}(\rho^q O_{\mathrm{A}})}{\mathbf{Tr}_{(\mathrm{A,B})}(\rho^q)}$	$\Longrightarrow$	$\dfrac{\mathbf{Tr}_{\mathrm{A}}(E_{\mathrm{A}}\mathcal{O}_{\mathrm{A}})}{\mathbf{Tr}_{\mathrm{A}}(E_{\mathrm{A}})}$	$\overset{\rho^q = \rho_{\mathrm{A}}^q \otimes \rho_{\mathrm{B}}^q \;\downarrow}{\Longrightarrow}$	$\dfrac{\mathbf{Tr}_{\mathrm{B}}(\rho_{\mathrm{B}}^q)\mathbf{Tr}_{\mathrm{A}}(\rho_{\mathrm{A}}^q\mathcal{O}_{\mathrm{A}})}{\mathbf{Tr}_{\mathrm{B}}(\rho_{\mathrm{B}}^q)\mathbf{Tr}_{\mathrm{A}}(\rho_{\mathrm{A}}^q)}$	$\Longrightarrow$	$\dfrac{\mathbf{Tr}_{\mathrm{A}}(\rho_{\mathrm{A}}^q\mathcal{O}_{\mathrm{A}})}{\mathbf{Tr}_{\mathrm{A}}(\rho_{\mathrm{A}}^q)}$
$\dfrac{\mathbf{Tr}_{(\mathrm{A,B})}(\rho^q O_{\mathrm{B}})}{\mathbf{Tr}_{(\mathrm{A,B})}(\rho^q)}$	$\Longrightarrow$	$\dfrac{\mathbf{Tr}_{\mathrm{B}}(E_{\mathrm{B}}\mathcal{O}_{\mathrm{B}})}{\mathbf{Tr}_{\mathrm{B}}(E_{\mathrm{B}})}$	$\Longrightarrow$	$\dfrac{\mathbf{Tr}_{\mathrm{A}}(\rho_{\mathrm{A}}^q)\mathbf{Tr}_{\mathrm{B}}(\rho_{\mathrm{B}}^q\mathcal{O}_{\mathrm{B}})}{\mathbf{Tr}_{\mathrm{A}}(\rho_{\mathrm{A}}^q)\mathbf{Tr}_{\mathrm{B}}(\rho_{\mathrm{B}}^q)}$	$\Longrightarrow$	$\dfrac{\mathbf{Tr}_{\mathrm{B}}(\rho_{\mathrm{B}}^q\mathcal{O}_{\mathrm{B}})}{\mathbf{Tr}_{\mathrm{B}}(\rho_{\mathrm{B}}^q)}$

Figure 3.1. For quantum mean values of any system A + B, a layout showing the summarized and formal derivation from the full Hilbert space to the corresponding Hilbert subspaces.

Concerning the aforementioned independence, for a compound ferromagnetic system A + B, Figure 3.2 is illustrative to clarify the formula for the q-statistical independence. This formula is a good approximation when q approaches 1; additionally, we notice kinks for low values of q, 0.1 and 0.4; these kinks indicate the formation of ferromagnetic clusters. In this figure, the spin values are $S_A = 2$ and $S_B =$

3/2, spins of Mn^{3+} and Mn^{4+}, respectively. Furthermore, in Figure 3.2 and the next two, we have used the mean-field model given by Eqs. (3.1) and (3.2) shown below.

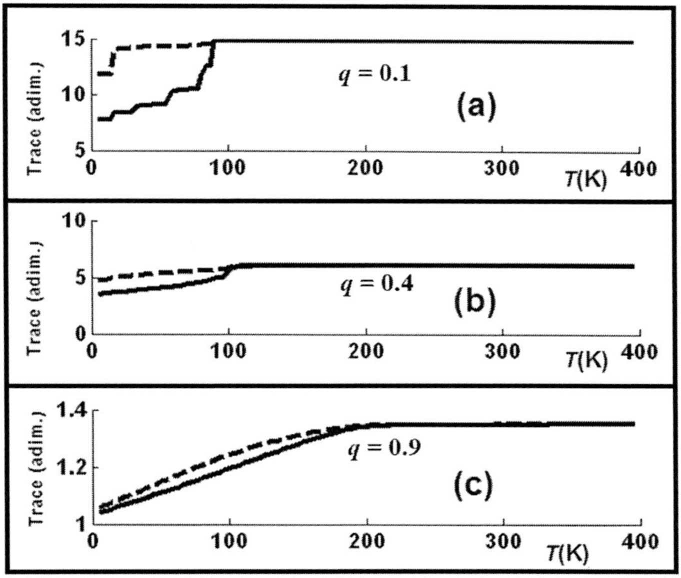

Figure 3.2. Graphics clarifying the formula $\rho^q = \rho_A^q \otimes \rho_B^q$. Solid lines represent $\mathbf{Tr}(\rho^q)$ vs. T; dashed lines, $\mathbf{Tr}(\rho_A^q \otimes \rho_B^q)$ vs. T. When q gets close to 1, the formula is a good approximation.

Figure 3.3 shows M vs. T for a ferromagnetic system A + B, with spins $S_A = 2$ and $S_B = 3/2$. Likewise, Figure 3.4 shows M vs. T for an antiferromagnetic system A + B, with spins $S_A = ½$ and $S_B = ½$. Clearly, with the model of the matrices E, the calculation of the mean values has no problem. But, surprisingly, in using the formulae of the third version of nonextensive statistical mechanics, there is no overlapping of the mean values calculated in different Hilbert spaces. Additionally, for $q = 0.1$, 0.3, 0.5, and 0.7, we again observe the formation of magnetic clusters, which is signaled by kinks. For the same physical system for Figure 3.3, Figure 3.5 shows that nonextensive magnetizations (and internal energies) become additive only with the model of the matrices E; for $q = 0.3$, we again observe kinks.

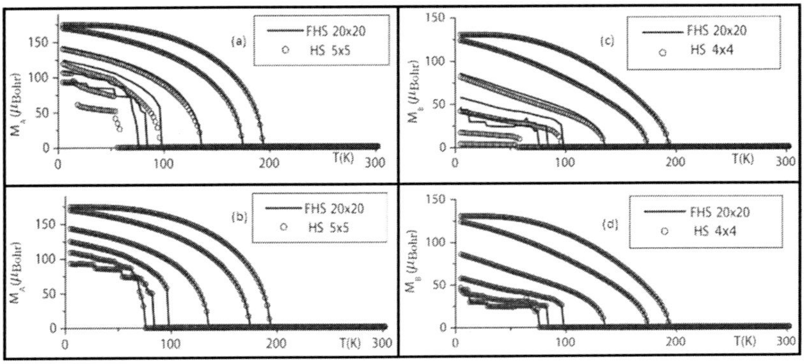

Figure 3.3. For a ferromagnet, $M_{A(B)}$ vs. T obtained by using the third version of nonextensive statistical mechanics (**a** and **c**). Also, the same parameters obtained by using the method of the matrices E (**b** and **d**). FHS means full Hilbert space; HS, Hilbert subspace. Besides, $S_A = 2$, $S_B = 3/2$, and $q = 0.1$, 0.3, 0.5, 0.7, 0.9, and 1.0.

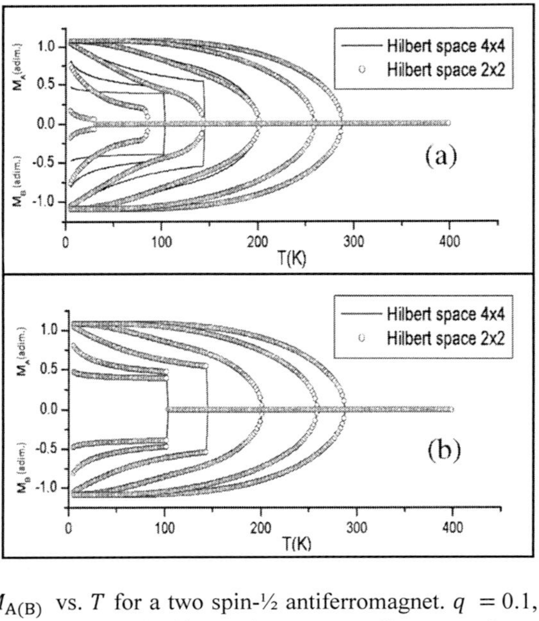

Figure 3.4. $M_{A(B)}$ vs. T for a two spin-½ antiferromagnet. $q = 0.1$, 0.3, 0.5, 0.7, 0.9, and 1.0. The bigger q, the bigger the corresponding saturation magnetization. We observe calculations done by using the third version of nonextensive statistical mechanics and the method of the matrices E (panels **a** and **b**, respectively.)

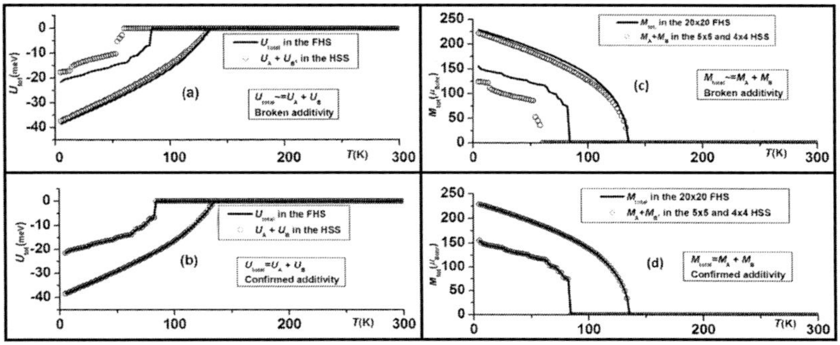

Figure 3.5. For a ferromagnet having spins $S_A = 2$ and $S_B = 3/2$, $U_{tot.}$ vs. T and $M_{tot.}$ vs. T obtained by using the third version of nonextensive statistical mechanics (**a** and **c**, respectively). We also have the same parameters obtained by using matrices E (**b** and **d**, respectively). The values of q are 0.3 and 0.7. The additivity of the two thermodynamic magnitudes is confirmed only by the method of the matrices E.

On the Exemplifying Magnetic Model to Be Utilized

Concretely, the utilized theoretical model is the Heisenberg model in the mean-field approximation (in 1907, the French physicist Pierre Ernest Weiss proposed the molecular-field approximation for ferromagnetism; this phenomenon and the origin of the Weiss field were explained by the German theoretical physicist Werner Karl Heisenberg in 1928). In this approach, we can replace all exchange interactions of the two crystal sublattices by two effective fields. Obviously, this causes the loss of some properties of the Heisenberg Hamiltonian, such as the short-range order at temperatures above the Curie temperature (in this Hamiltonian, the spins are quantum as opposed to, e.g., the Ising model which considers classical spins).

In the scientific literature, the mean-field approximation is known to predict temperatures higher than the actual critical values. However, the use of a mean-field theory is justified since a solid has a number of atoms close to Avogadro's number; thus, infinite degrees of freedom are considered. Although using a mean-field approximation means disregarding magnetic fluctuations, we use this theory as a guide to studying manganites. Even though these compounds are strongly correlated systems, several publications show that the use of a mean-field model gives good results when investigating manganites (e.g. Reis Jr. 2003).

Next, we have the magnetic Hamiltonian ($g = 2$ for a spin-½):

$$H = -2\mu_{\text{Bohr}} S_z^{\text{A}} B_{\text{A}} - 2\mu_{\text{Bohr}} S_z^{\text{B}} B_{\text{B}} \,, \qquad (3.1)$$

where B_{A} and B_{B} are the effective magnetic fields acting on the respective systems A and B, that is, magnetic sublattices A and B; S_z^{A} and S_z^{B} are the matrix representations of the spin operators in the respective subsystems. In the mean-field approximation, these magnetic fields are given by

$$B_{\text{A(B)}} = B_0 + \lambda_{\text{A(B)}} M_{\text{A(B)}} + \lambda_{\text{AB}} M_{\text{B(A)}} \,, \qquad (3.2)$$

$\lambda_{\text{A(B)}}$ represents the corresponding intra-lattice coupling, while λ_{AB} symbolizes the interlattice coupling; $\lambda_{\text{AB}} > 0$ means a ferromagnetic coupling and $\lambda_{\text{AB}} < 0$, an antiferromagnetic coupling. The spin operators S_z^{A} and S_z^{B} are dimensionless, for ħ is contained in the Bohr magneton μ_{Bohr}. The previously mentioned Figures 3.2, 3.3, 3.4 and 3.5 are computer calculations done by using this exemplifying model.

In the following subsections, we have two analyzes explaining why nonextensive magnetization curves do not overlap when calculated in different Hilbert spaces.

Mean Values of a Two Spin-½ System (Third Version of Nonextensive Statistical Mechanics)

Firstly, it must be pointed out that the detailed calculations for this subsection, and the following two, can be revised in the next section.

Thus, for an observable of a compound system A + B, we will see that the third version of nonextensive statistical mechanics leads to dissimilar mean values when calculated in different Hilbert spaces. By using the density matrix formalism, below, we will carry out calculations for the magnetic Hamiltonian from Eq. (3.1). Then, the density matrix of the system is given by

$$\rho = \frac{1}{Z_q}\left[\mathbb{1} + (1-q)\frac{2\mu_{\text{Bohr}}}{kT}\left(S_z^{\text{A}} B_{\text{A}} + S_z^{\text{B}} B_{\text{B}}\right)\right]^{\frac{1}{1-q}} \,, \qquad (3.3)$$

where $\mathbb{1}$ is the identity matrix in the 4x4 Hilbert space and Z_q, the partition function

$$Z_q = \mathbf{Tr}\left\{\left[\mathbb{1} + (1-q)\frac{2\mu_{\text{Bohr}}}{kT}\left(S_z^{\text{A}} B_{\text{A}} + S_z^{\text{B}} B_{\text{B}}\right)\right]^{\frac{1}{1-q}}\right\} . \qquad (3.4)$$

To find the spin matrices in the 4x4 full Hilbert space, $S_z^{\text{A(B)}}$, we will need the spin matrices in the 2x2 Hilbert subspaces, $s_z^{\text{A(B)}}$, because a two spin-½ system satisfies these formulae:

$$S_z^{\text{A}} = \frac{1}{\mathbf{Tr}_{\text{B}}(\mathbb{1}_{\text{B}})} s_z^{\text{A}} \otimes \mathbb{1}_{\text{B}} \quad \text{and} \quad S_z^{\text{B}} = \frac{1}{\mathbf{Tr}_{\text{A}}(\mathbb{1}_{\text{A}})} \mathbb{1}_{\text{A}} \otimes s_z^{\text{B}} \,, \qquad (3.5)$$

where $\mathbb{1}_{\mathrm{A}}$ and $\mathbb{1}_{\mathrm{B}}$ are the identity matrices in the 2x2 Hilbert subspaces. $\mathbf{Tr}_{\mathrm{A}}(\mathbb{1}_{\mathrm{A}})$ and $\mathbf{Tr}_{\mathrm{B}}(\mathbb{1}_{\mathrm{B}})$ are the normalization factors; consequently, we can obtain $\mathbf{Tr}_{\mathrm{A}}(S_z^{\mathrm{B}}) = \mathrm{s}_z^{\mathrm{B}}$ and $\mathbf{Tr}_{\mathrm{B}}(S_z^{\mathrm{A}}) = \mathrm{s}_z^{\mathrm{A}}$.

In the 2x2 Hilbert subspaces, the spin matrices are given by the following expressions:

$$\mathrm{s}_z^{(\mathrm{A,B})} = \frac{1}{2}\begin{pmatrix} 1 & 0 \\ 0 & -1 \end{pmatrix}, \qquad (3.6)$$

and in the 4x4 full Hilbert space, by these ones:

$$\mathrm{S}_z^{\mathrm{A}} = \frac{1}{4}\begin{pmatrix} 1 & 0 & 0 & 0 \\ 0 & 1 & 0 & 0 \\ 0 & 0 & -1 & 0 \\ 0 & 0 & 0 & -1 \end{pmatrix} \text{ and } \mathrm{S}_z^{\mathrm{B}} = \frac{1}{4}\begin{pmatrix} 1 & 0 & 0 & 0 \\ 0 & -1 & 0 & 0 \\ 0 & 0 & 1 & 0 \\ 0 & 0 & 0 & -1 \end{pmatrix}. \qquad (3.7)$$

Therefore, using Eq. (3.3), matrix ρ is given as follows:

$$\rho = \frac{1}{Z_q}\begin{pmatrix} \alpha_1 & 0 & 0 & 0 \\ 0 & \alpha_2 & 0 & 0 \\ 0 & 0 & \alpha_3 & 0 \\ 0 & 0 & 0 & \alpha_4 \end{pmatrix}, \qquad (3.8)$$

where we have entered these parameters:

$$\alpha_1 = \left[1 + (1-q)\frac{\mu_{\mathrm{Bohr}}}{2kT}(B_{\mathrm{A}} + B_{\mathrm{B}})\right]^{\frac{1}{1-q}}, \quad \alpha_2 = \left[1 + (1-q)\frac{\mu_{\mathrm{Bohr}}}{2kT}(B_{\mathrm{A}} - B_{\mathrm{B}})\right]^{\frac{1}{1-q}},$$

$$\alpha_3 = \left[1 + (1-q)\frac{\mu_{\mathrm{Bohr}}}{2kT}(-B_{\mathrm{A}} + B_{\mathrm{B}})\right]^{\frac{1}{1-q}}, \text{and } \alpha_4 = \left[1 - (1-q)\frac{\mu_{\mathrm{Bohr}}}{2kT}(B_{\mathrm{A}} + B_{\mathrm{B}})\right]^{\frac{1}{1-q}}, \quad (3.9)$$

noticeably, the partition function is $Z_q = \alpha_1 + \alpha_2 + \alpha_3 + \alpha_4$. Thus, raising ρ to the power q, we have

$$\rho^q = \frac{1}{Z_q^q}\begin{pmatrix} \alpha_1^q & 0 & 0 & 0 \\ 0 & \alpha_2^q & 0 & 0 \\ 0 & 0 & \alpha_3^q & 0 \\ 0 & 0 & 0 & \alpha_4^q \end{pmatrix}. \qquad (3.10)$$

In short, using Eqs. (2.65a), (3.7), and (3.10), in the 4x4 Hilbert space, the nonextensive magnetization of the subsystem A is given by

$$M_z^{\mathrm{A}} = \frac{\alpha_1^q + \alpha_2^q - \alpha_3^q - \alpha_4^q}{\alpha_1^q + \alpha_2^q + \alpha_3^q + \alpha_4^q}\frac{\mu_{\mathrm{Bohr}}}{m_{\mathrm{A}}}; \qquad (3.11)$$

for system B, the calculations are similar.

Likewise, we now summarize how to calculate the nonextensive magnetization in the 2x2 Hilbert subspaces: To use the formulae of the third version of nonextensive statistical mechanics, from Eq. (3.8) we take the partial trace over the states of the system B, thus obtaining the following matrix ρ_A:

$$\rho_{\mathrm{A}} = \frac{1}{Z_q^q}\begin{pmatrix} \alpha_1 + \alpha_2 & 0 \\ 0 & \alpha_3 + \alpha_4 \end{pmatrix}, \qquad (3.12)$$

and raising it to the power q, we obtain the following expression:

$$\rho_{\mathrm{A}}^{q} = \frac{1}{z_q^q}\begin{pmatrix}(\alpha_1+\alpha_2)^q & 0\\ 0 & (\alpha_3+\alpha_4)^q\end{pmatrix}; \qquad (3.13)$$

therefore, by using Eqs. (2.66a), (3.6), and (3.13), in the 2x2 Hilbert subspace, the nonextensive magnetization of the subsystem A is expressed as follows:

$$M_z^{\mathrm{A}} = \frac{(\alpha_1+\alpha_2)^q - (\alpha_3+\alpha_4)^q}{(\alpha_1+\alpha_2)^q + (\alpha_3+\alpha_4)^q}\frac{\mu_{\mathrm{Bohr}}}{m_{\mathrm{A}}}. \qquad (3.14)$$

This equation is *different* from Eq. (3.11). Besides, in Eq. (2.66a), $\mathrm{j}_z^{\mathrm{A}} \equiv \mathrm{s}_z^{\mathrm{A}}$ has been considered.

Applying an Alternative Method to Calculate Mean Values of Compound Systems

There is another method to compute mean values. This one arises when considering the equivalence from Eq. (2.63) ($\langle\hat{\mathcal{O}}_{\mathrm{A}}\rangle_{\mathrm{FHS}} \equiv \langle\hat{\mathcal{O}}_{\mathrm{A}}\rangle_{\mathrm{HS}}$ and $\langle\hat{\mathcal{O}}_{\mathrm{B}}\rangle_{\mathrm{FHS}} \equiv \langle\hat{\mathcal{O}}_{\mathrm{B}}\rangle_{\mathrm{HS}}$); this is how matrices E_{A} and E_{B} appear. Matrices E_{A} and E_{B} are different from ρ_{A} and ρ_{B}, respectively, since they are obtained by taking the respective partial trace of ρ^q, i.e., we will take the partial trace of the power of a matrix. The usual method consists in taking the partial trace of a matrix and, afterward, raising it to the power q. Next, we establish the definition of the matrices E in an equation numbered:

$$E_{\mathrm{A}} = \mathbf{Tr}_{\mathrm{B}}\,(\rho^q) \qquad \text{and} \qquad E_{\mathrm{B}} = \mathbf{Tr}_{\mathrm{A}}\,(\rho^q) \qquad (3.15)$$

where ρ^q is the density matrix of the system $\mathrm{A}+\mathrm{B}$, which is raised explicitly to the power q.

Therefore, for subsystems A and B, in the respective Hilbert subspaces, the mean values of any observable are defined in this manner:

$$\langle\hat{\mathcal{O}}_{\mathrm{A}}\rangle = \frac{\mathbf{Tr}_{\mathrm{A}}(E_{\mathrm{A}}\mathcal{O}_{\mathrm{A}})}{\mathbf{Tr}_{\mathrm{A}}(E_{\mathrm{A}})}, \qquad (3.16a)$$

$$\langle\hat{\mathcal{O}}_{\mathrm{B}}\rangle = \frac{\mathbf{Tr}_{\mathrm{B}}(E_{\mathrm{B}}\mathcal{O}_{\mathrm{B}})}{\mathbf{Tr}_{\mathrm{B}}(E_{\mathrm{B}})}, \qquad (3.16b)$$

where $\hat{\mathcal{O}}_{\mathrm{A}}$ and $\hat{\mathcal{O}}_{\mathrm{B}}$ are operators acting in subspaces A and B, respectively; obviously, $\mathcal{O}_{\mathrm{A}}$ and $\mathcal{O}_B$ are the matrices representing them. For example, the magnetizations are defined thus:

$$M_{\mathrm{A}} = \frac{\mathbf{Tr}_{\mathrm{A}}(g_{\mathrm{A}}E_{\mathrm{A}}\mathrm{s}_z^{\mathrm{A}})}{\mathbf{Tr}_{\mathrm{A}}(E_{\mathrm{A}})}\frac{\mu_{\mathrm{Bohr}}}{m_{\mathrm{A}}}, \qquad (3.17a)$$

$$M_{\mathrm{B}} = \frac{\mathbf{Tr}_{\mathrm{B}}(g_{\mathrm{B}}E_{\mathrm{B}}\mathrm{s}_z^{\mathrm{B}})}{\mathbf{Tr}_{\mathrm{B}}(E_{\mathrm{B}})}\frac{\mu_{\mathrm{Bohr}}}{m_{\mathrm{B}}}. \qquad (3.17b)$$

To apply this new method to the system represented by Eq. (3.1), we must obtain E_{A} and E_{B} from Eq. (3.10). Concerning the former, in Eq.

(3.10), we take the partial trace over the states of the system B and get as follows:

$$E_{\mathrm{A}} = \frac{1}{Z_q^q}\begin{pmatrix} \alpha_1^q + \alpha_2^q & 0 \\ 0 & \alpha_3^q + \alpha_4^q \end{pmatrix}. \tag{3.18}$$

In summary, in the 2x2 Hilbert subspace, by using Eqs. (3.6), (3.17a), and (3.18), we obtain the nonextensive magnetization of the subsystem A:

$$M_z^{\mathrm{A}} = \frac{\alpha_1^q + \alpha_2^q - \alpha_3^q - \alpha_4^q}{\alpha_1^q + \alpha_2^q + \alpha_3^q + \alpha_4^q}\frac{\mu_{\mathrm{Bohr}}}{m_{\mathrm{A}}}. \tag{3.19}$$

This result is *consistent* with Eq. (3.11). For subsystem B, the calculations are completely analogous.

Additional Calculation for a System A + B With Interaction

Let consider two interacting spin-½ systems, A and B, with the following Hamiltonian:

$$H = -2\mu_{\mathrm{Bohr}}(\mathrm{S}_z^{\mathrm{A}} + \mathrm{S}_z^{\mathrm{B}})B_0 - \xi \mathrm{S}_z^{\mathrm{A}}\mathrm{S}_z^{\mathrm{B}}, \tag{3.20}$$

where ξ is the coupling parameter of the interacting systems and B_0, the external magnetic field. Accordingly, in the third version of nonextensive statistical mechanics, the density matrix ρ is given in this manner:

$$\rho = \frac{1}{Z_q}\left\{\mathbb{1} + (1-q)\frac{2\mu_{\mathrm{Bohr}}(\mathrm{S}_z^{\mathrm{A}} + \mathrm{S}_z^{\mathrm{B}})B_0 + \xi \mathrm{S}_z^{\mathrm{A}}\mathrm{S}_z^{\mathrm{B}}}{kT}\right\}^{\frac{1}{1-q}}. \tag{3.21}$$

Then, by using $\mathrm{S}_z^{\mathrm{A}}$ and $\mathrm{S}_z^{\mathrm{B}}$ from Eqs. (3.7), ρ is given by the following expression:

$$\rho = \frac{1}{Z_q}\begin{pmatrix} \alpha_1 & 0 & 0 & 0 \\ 0 & \beta_1 & 0 & 0 \\ 0 & 0 & \beta_1 & 0 \\ 0 & 0 & 0 & \gamma_1 \end{pmatrix}, \tag{3.22}$$

with the following parameters:

$$\alpha_1 = \left[1 + (1-q)\frac{\mu_{\mathrm{Bohr}}B_0 + {}^{\xi}/_{8}}{kT}\right]^{\frac{1}{1-q}}, \quad \beta_1 = \left[1 + (1-q)\frac{-{}^{\xi}/_{8}}{kT}\right]^{\frac{1}{1-q}}, \text{ and}$$

$$\gamma_1 = \left[1 + (1-q)\frac{-\mu_{\mathrm{Bohr}}B_0 + {}^{\xi}/_{8}}{kT}\right]^{\frac{1}{1-q}}, \tag{3.23}$$

raising to q, we have as follows:

$$\rho^q = \frac{1}{Z_q^q}\begin{pmatrix} \alpha_1^q & 0 & 0 & 0 \\ 0 & \beta_1^q & 0 & 0 \\ 0 & 0 & \beta_1^q & 0 \\ 0 & 0 & 0 & \gamma_1^q \end{pmatrix}. \tag{3.24}$$

In short, with the use of Eqs. (2.65a), (3.7), and (3.24), in the 4x4 Hilbert space, the magnetization of the subsystem A becomes thus:

$$M_z^{\mathrm{A}} = \frac{\alpha_1^q - \gamma_1^q}{\alpha_1^q + 2\beta_1^q + \gamma_1^q}\frac{\mu_{\mathrm{Bohr}}}{m_{\mathrm{A}}}. \tag{3.25}$$

Meanwhile, to use the third version of nonextensive statistical mechanics in the 2x2 Hilbert space, from Eq. (3.22) we obtain the following partial matrix:

$$\rho_{\mathrm{A}} = \frac{1}{Z_q}\begin{pmatrix} \alpha_1 + \beta_1 & 0 \\ 0 & \beta_1 + \gamma_1 \end{pmatrix}, \tag{3.26}$$

raising to q

$$\rho_{\mathrm{A}}^q = \frac{1}{Z_q^q}\begin{pmatrix} (\alpha_1 + \beta_1)^q & 0 \\ 0 & (\beta_1 + \gamma_1)^q \end{pmatrix}. \tag{3.27}$$

Therefore, in the 2x2 Hilbert subspace, by using Eqs. (2.66a), (3.6), and (3.27), the nonextensive magnetization of the subsystem A is given by

$$M_z^{\mathrm{A}} = \frac{(\alpha_1 + \beta_1)^q - (\beta_1 + \gamma_1)^q}{(\alpha_1 + \beta_1)^q + (\beta_1 + \gamma_1)^q}\frac{\mu_{\mathrm{Bohr}}}{m_{\mathrm{A}}}. \tag{3.28}$$

Besides, in Eq. (2.66a), $\mathrm{j}_z^A \equiv \mathrm{s}_z^A$ has been considered. We can verify that Eq. (3.28) is different from Eq. (3.25).

Furthermore, we now will use the method of the matrices E; we apply it to the system represented by Eq. (3.20); in Eq. (3.24), we take the partial trace over the states of the system B. Thus, we obtain the following expression:

$$E_{\mathrm{A}} = \frac{1}{Z_q^q}\begin{pmatrix} \alpha_1^q + \beta_1^q & 0 \\ 0 & \beta_1^q + \gamma_1^q \end{pmatrix}. \tag{3.29}$$

In short, in the 2x2 Hilbert subspace, by means of Eqs. (3.6), (3.17a), and (3.29), the nonextensive magnetization of the subsystem A is given as follows:

$$M_z^{\mathrm{A}} = \frac{\alpha_1^q - \gamma_1^q}{\alpha_1^q + 2\beta_1^q + \gamma_1^q}\frac{\mu_{\mathrm{Bohr}}}{m_{\mathrm{A}}}. \tag{3.30}$$

This result is *consistent* with Eq. (3.25).

Some Detailed Calculations

Before showing the calculations, it must be indicated some basic concepts and formulae to be used. Firstly, in quantum mechanics, the physical system state is identified by a vector in a complex Hilbert space: $|u\rangle$. Each vector is termed a ket and associated with it, we have a dual vector named bra: $\langle u|$. It is also defined the outer product of $|u\rangle$ and $\langle v|$: $|u\rangle\langle v|$. If we consider a two spin-½ system, for each subsystem in its corresponding subspace we have $|u\rangle = |\pm\rangle$ and $\langle v| = \langle\pm|$, where $|+\rangle$ means a spin-up state and $|-\rangle$, a spin-down state; the same for the dual vectors.

To find the spin operators in the 4x4 full Hilbert space, $\hat{\mathrm{S}}^z_{\mathrm{A(B)}}$, we need the spin operators in the 2x2 Hilbert subspaces, $\hat{\mathrm{s}}^z_{\mathrm{A(B)}}$. Then, for a two spin-½ system we have the following expressions:

$$\hat{\mathrm{S}}^z_{\mathrm{A}} = \frac{1}{\mathbf{Tr}_{\mathrm{B}}(\hat{\mathbb{1}}_{\mathrm{B}})} \hat{\mathrm{s}}^z_{\mathrm{A}} \otimes \hat{\mathbb{1}}_B \text{ and } \hat{\mathrm{S}}^z_{\mathrm{B}} = \frac{1}{\mathbf{Tr}_{\mathrm{A}}(\hat{\mathbb{1}}_{\mathrm{A}})} \hat{\mathbb{1}}_{\mathrm{A}} \otimes \hat{\mathrm{s}}^z_{\mathrm{B}}\,, \quad (3.31)$$

where $\hat{\mathbb{1}}_{\mathrm{A}}$ and $\hat{\mathbb{1}}_{\mathrm{B}}$ are the identity operators in the 2x2 Hilbert subspaces. $\mathbf{Tr}_{\mathrm{A}}(\hat{\mathbb{1}}_{\mathrm{A}})$ and $\mathbf{Tr}_{\mathrm{B}}(\hat{\mathbb{1}}_{\mathrm{B}})$ are normalization factors so that $\mathbf{Tr}_{\mathrm{A}}(\hat{\mathrm{S}}^z_{\mathrm{B}}) = \hat{\mathrm{s}}^z_{\mathrm{B}}$ and $\mathbf{Tr}_{\mathrm{B}}(\hat{\mathrm{S}}^z_{\mathrm{A}}) = \hat{\mathrm{s}}^z_{\mathrm{A}}$. Besides, the identity operators, and their corresponding traces, in the 2x2 Hilbert subspaces are given as follows:

$$\hat{\mathbb{1}}_{\mathrm{A(B)}} = |+\rangle\langle+| + |-\rangle\langle-| \quad \text{and} \quad \mathbf{Tr}_{\mathrm{A(B)}}(\hat{\mathbb{1}}_{\mathrm{A(B)}}) = 2\,. \quad (3.32)$$

By considering Eqs. (3.5) as well as the following expressions of the spin operators in the 2x2 Hilbert subspaces:

$$\hat{\mathrm{s}}^z_{\mathrm{A(B)}} = \frac{1}{2}\{|+\rangle\langle+| - |-\rangle\langle-|\}\,, \quad (3.33)$$

we arrive at the spin operators in the 4x4 full Hilbert space:

$$\hat{\mathrm{S}}^z_{\mathrm{A}} = \frac{1}{4}\{\,|+\,+\rangle\langle+\,+| + |+\,-\rangle\langle+\,-| - |-\,+\rangle\langle-\,+| - |-\,-\rangle\langle-\,-|\,\}\,, \quad (3.34a)$$

$$\hat{\mathrm{S}}^z_{\mathrm{B}} = \frac{1}{4}\{\,|+\,+\rangle\langle+\,+| - |+\,-\rangle\langle+\,-| + |-\,+\rangle\langle-\,+| - |-\,-\rangle\langle-\,-|\,\}\,, \quad (3.34b)$$

where $|\pm\,\pm\rangle$ stand for the compound-system states, e.g., $|+\,-\rangle$ represents the subsystem A with spin up and subsystem B with spin down.

Besides, in the 4x4 full Hilbert space, the matrix representations of the identity and spin operators are, respectively,

$$\mathbb{1}_{\mathrm{A(B)}} = \begin{pmatrix} 1 & 0 & 0 & 0 \\ 0 & 1 & 0 & 0 \\ 0 & 0 & 1 & 0 \\ 0 & 0 & 0 & 1 \end{pmatrix}, \quad (3.35a)$$

$$\mathrm{S}_\mathrm{A}^z = \frac{1}{4}\begin{pmatrix} 1 & 0 & 0 & 0 \\ 0 & 1 & 0 & 0 \\ 0 & 0 & -1 & 0 \\ 0 & 0 & 0 & -1 \end{pmatrix} \text{ and } \mathrm{S}_\mathrm{B}^z = \frac{1}{4}\begin{pmatrix} 1 & 0 & 0 & 0 \\ 0 & -1 & 0 & 0 \\ 0 & 0 & 1 & 0 \\ 0 & 0 & 0 & -1 \end{pmatrix}. \quad (3.35b)$$

In the Hilbert subspaces, the representations of the previously mentioned operators are, respectively,

$$\mathbb{1}_{\mathrm{A(B)}} = \begin{pmatrix} 1 & 0 \\ 0 & 1 \end{pmatrix}, \quad (3.36a)$$

$$\mathrm{s}_{\mathrm{A(B)}}^z = \frac{1}{2}\begin{pmatrix} 1 & 0 \\ 0 & -1 \end{pmatrix}. \quad (3.36b)$$

Analytical Calculus for q = 2

For teaching purposes, below, we have specific calculations for the same physical system that comply with Eq. (3.20) in the subextensive case $q = 2$. For clarity, from now on, we will call it Eq. (3.37):

$$\widehat{H} = -2\mu_{\mathrm{Bohr}}(\hat{\mathrm{S}}_z^\mathrm{A} + \hat{\mathrm{S}}_z^\mathrm{B})B_0 - \xi\hat{\mathrm{S}}_z^\mathrm{A}\hat{\mathrm{S}}_z^\mathrm{B}, \quad (3.37)$$

where ξ is the coupling parameter and B_0, the external magnetic field. In accordance with nonextensive statistical mechanics, the density operator is given as follows:

$$\hat{\rho} = \frac{1}{Z_q}\left\{\hat{\mathbb{1}} + (1-q)\frac{2\mu_{\mathrm{Bohr}}(\hat{\mathrm{S}}_z^\mathrm{A} + \hat{\mathrm{S}}_z^\mathrm{B})B_0 + \xi\hat{\mathrm{S}}_z^\mathrm{A}\hat{\mathrm{S}}_z^\mathrm{B}}{kT}\right\}^{\frac{1}{1-q}}, \quad (3.38)$$

evaluating at $q = 2$ we have

$$\hat{\rho} = \frac{1}{Z_2}\left\{\hat{\mathbb{1}} - \frac{2\mu_{\mathrm{Bohr}}(\hat{\mathrm{S}}_z^\mathrm{A} + \hat{\mathrm{S}}_z^\mathrm{B})B_0 + \xi\hat{\mathrm{S}}_z^\mathrm{A}\hat{\mathrm{S}}_z^\mathrm{B}}{kT}\right\}^{-1}, \quad (3.39)$$

where Z_2 is the following partition function:

$$Z_2 = \mathbf{Tr}\left[\left\{\hat{\mathbb{1}} - \frac{2\mu_{\mathrm{Bohr}}(\hat{\mathrm{S}}_z^\mathrm{A} + \hat{\mathrm{S}}_z^\mathrm{B})B_0 + \xi\hat{\mathrm{S}}_z^\mathrm{A}\hat{\mathrm{S}}_z^\mathrm{B}}{kT}\right\}^{-1}\right]. \quad (3.40)$$

Consequently, by using Eqs. (3.34a) and (3.34b), the density operator is given by

$$\hat{\rho} = \frac{1}{Z_2}\{\alpha_1|++\rangle\langle++| + \beta_1|+-\rangle\langle+-| + \beta_1|-+\rangle\langle-+| - \gamma_1|--\rangle\langle--|\}, \quad (3.41)$$

where we have defined the corresponding new parameters this way:

$$\alpha_1 = \left[1 - \frac{\mu_{\mathrm{Bohr}}B_0 + \xi/8}{kT}\right]^{-1}, \beta_1 = \left[1 + \frac{\xi/8}{kT}\right]^{-1}, \text{and } \gamma_1 = \left[1 - \frac{-\mu_{\mathrm{Bohr}}B_0 + \xi/8}{kT}\right]^{-1}. \quad (3.42)$$

Thus, the density operator raised to the second power is given as follows:

$$\hat{\rho}^2 = \frac{1}{Z_2^2}\{\alpha_1^2|++\rangle\langle++| + \beta_1^2|+-\rangle\langle+-| + \beta_1^2|-+\rangle\langle-+| + \gamma_1^2|--\rangle\langle--|\}; \quad (3.43)$$

in a matrix formulation, $\hat{\rho}$ and $\hat{\rho}^2$ are, respectively, represented by the following expressions:

$$\rho = \frac{1}{Z_2}\begin{pmatrix} \alpha_1 & 0 & 0 & 0 \\ 0 & \beta_1 & 0 & 0 \\ 0 & 0 & \beta_1 & 0 \\ 0 & 0 & 0 & \gamma_1 \end{pmatrix}, \qquad (3.44)$$

and

$$\rho^2 = \frac{1}{Z_2^2}\begin{pmatrix} \alpha_1^2 & 0 & 0 & 0 \\ 0 & \beta_1^2 & 0 & 0 \\ 0 & 0 & \beta_1^2 & 0 \\ 0 & 0 & 0 & \gamma_1^2 \end{pmatrix}. \qquad (3.45)$$

For the subsystem A in the 4x4 full Hilbert space, by using Eqs. (2.65a), (3.34a), and (3.45), the nonextensive magnetization is given as follows:

$$M_z^{\mathrm{A}} = \frac{\alpha_1^2 - \gamma_1^2}{\alpha_1^2 + 2\beta_1^2 + \gamma_1^2}\frac{\mu_{\mathrm{Bohr}}}{m_{\mathrm{A}}}. \qquad (3.46)$$

It must also be indicated that, in Eq. (2.65a), $J_z^{\mathrm{A}} \equiv \mathrm{S}_z^{\mathrm{A}}$ has been considered.

Meanwhile, to use nonextensive statistical mechanics in the 2x2 Hilbert subspace, from Eq. (3.41), we must first obtain $\hat{\rho}_{\mathrm{A}}$. Thus, by applying the partial trace operation to that equation, we have the following expression:

$$\hat{\rho}_{\mathrm{A}} = \frac{1}{Z_2}\{(\alpha_1 + \beta_1)|+\rangle\langle+| + (\beta_1 + \gamma_1)\,|-\rangle\langle-|\} \qquad (3.47)$$

and then we square it

$$\hat{\rho}_{\mathrm{A}}^2 = \frac{1}{Z_2}\{(\alpha_1 + \beta_1)^2|+\rangle\langle+| + (\beta_1 + \gamma_1)^2|-\rangle\langle-|\}. \qquad (3.48)$$

In a matrix representation, $\hat{\rho}_A$ and $\hat{\rho}_A^2$ are, respectively,

$$\rho_{\mathrm{A}} = \frac{1}{Z_2}\begin{pmatrix} \alpha_1 + \beta_1 & 0 \\ 0 & \beta_1 + \gamma_1 \end{pmatrix}, \qquad (3.49a)$$

$$\rho_{\mathrm{A}}^2 = \frac{1}{Z_2^2}\begin{pmatrix} (\alpha_1 + \beta_1)^2 & 0 \\ 0 & (\beta_1 + \gamma_1)^2 \end{pmatrix}. \qquad (3.49b)$$

Therefore, for the subsystem A in the 2x2 Hilbert subspace, by means of Eqs. (2.66a), (3.36b), and (3.49b), the nonextensive magnetization is given as follows:

$$M_z^{\mathrm{A}} = \frac{(\alpha_1 + \beta_1)^2 - (\beta_1 + \gamma_1)^2}{(\alpha_1 + \beta_1)^2 + 2}\frac{\mu_{\mathrm{Bohr}}}{m_{\mathrm{A}}}, \qquad (3.50)$$

besides, in Eq. (2.66a), $\mathrm{j}_z^{\mathrm{A}} \equiv \mathrm{s}_{\mathrm{A(B)}}^z$ has been considered. We realize Eq. (3.50) is different from Eq. (3.46).

On the other hand, now we will use the method of the matrices E. Thus, in Eq. (3.43), we take the partial trace over the states B, obtaining this expression:

$$E_{\mathrm{A}} = \frac{1}{Z_2^2}\{(\alpha_1^2+\beta_1^2)|+\rangle\langle+| + (\beta_1^2+\gamma_1^2)|-\rangle\langle-|\}, \quad (3.51)$$

which can be represented in a matrix array:

$$E_{\mathrm{A}} = \frac{1}{Z_2^2}\begin{pmatrix} \alpha_1^2+\beta_1^2 & 0 \\ 0 & \beta_1^2+\gamma_1^2 \end{pmatrix}. \quad (3.52)$$

Then, for the subsystem A in the 2x2 Hilbert subspace, by using Eqs. (3.17a), (3.36b), and (3.52), the nonextensive magnetization is given as follows:

$$M_z^{\mathrm{A}} = \frac{\alpha_1^2-\gamma_1^2}{\alpha_1^2+2\beta_1^2+\gamma_1^2}\,\frac{\mu_{\mathrm{Bohr}}}{m_{\mathrm{A}}}. \quad (3.53)$$

This result is consistent with Eq. (3.46). For subsystem B, the calculations are similar.

The Escort Distributions of Compound Systems in Nonextensive Statistical Mechanics

If the escort distributions, introduced in Eq. (2.70), are used in single systems, no problem appears when calculating quantum mean values; however, whenever compound systems are studied, the picture changes.

Thus, by using the aforementioned distributions, a derivation from the FHS mean values toward the HS mean values is schematized in Figure 3.6, where the escort distributions P and P_{A} have already been defined in Eqs. (2.70) and (2.73a), respectively (for subsystem B, the calculations are analogous). Thereupon, O_{A} is the matrix representation of the operator $\hat{O}_{\mathrm{A}}$ (in the full Hilbert space), which acts on states A; $\mathcal{O}_{\mathrm{A}}$ is the analogous matrix for HS operator $\hat{\mathcal{O}}_{\mathrm{A}}$, which acts on the same states.

In the previously mentioned figure, Step **1** represents the quantum mean value of the observable O_{A} in the full Hilbert space. In the first line, we have the definition of the mean value according to the formulation based on the escort distributions; in the second line, we have the definition of the mean value in conformity with the third version of nonextensive statistical mechanics. The escort distributions ensure that both formulae of step **1** are equivalent. For the second line, steps **2**, **3**, and **4** are equivalent to steps **II**, **III**, and **IV** from the boxed formulae in Figure 3.1. It should be noted that to go from step **2** to step **3**, we must consider the statistical independence of the systems A and B. Therefore, the two final formulae from step **4** would have to be equivalent (but matrices E indicate that they are not), which implies

$$P_{\text{A}} = \frac{\rho_{\text{A}}^{q}}{\mathbf{Tr}_{\text{A}}(\rho_{\text{A}}^{q})} \quad \text{and} \quad P_{\text{B}} = \frac{\rho_{\text{B}}^{q}}{\mathbf{Tr}_{\text{B}}(\rho_{\text{B}}^{q})}\,. \qquad (3.54)$$

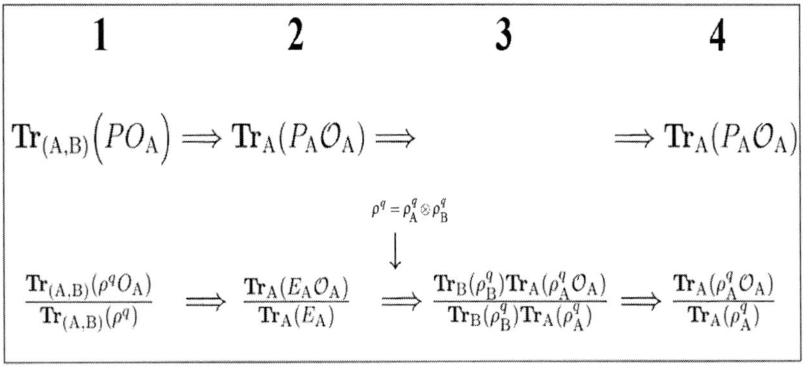

Figure 3.6. For quantum mean values and escort distributions, a layout showing the (summarized) formal derivation from the full Hilbert space to the corresponding Hilbert subspace. It is valid for any compound system A + B.

Figure 3.7 shows that the escort distributions do not apply to compound systems. The simulation is valid for a ferromagnetic system A + B, with spins $S_{\text{A}} = ½$ and $S_{\text{B}} = ½$; we have used the Heisenberg model in the mean-field approximation. Finally, we have two relationships that are a consequence of the method of the matrices E:

$$P_{\text{A}} = \frac{E_{\text{A}}}{\mathbf{Tr}_{\text{A}}(E_{\text{A}})} \quad \text{and} \quad P_{\text{B}} = \frac{E_{\text{B}}}{\mathbf{Tr}_{\text{B}}(E_{\text{B}})}\,. \qquad (3.55)$$

Clearly, formulas from (3.54) and (3.55) are not equal, respectively. The analysis introduced, aiming at experimental applications, clarifies the important issue of partial matrices in the escort distributions.

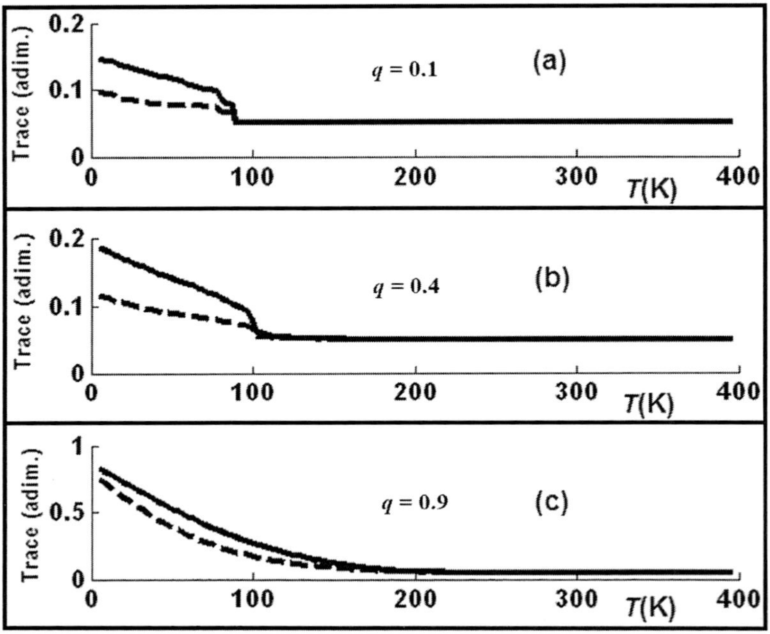

Figure 3.7. The distributions escort do not apply to a compound system A + B. The solid lines represent $\mathbf{Tr}_{\mathrm{A}}\left(P_{\mathrm{A}}^{2}\right)$; the dashed lines, $\mathbf{Tr}\left\{\left[\rho_{\mathrm{A}}^{q}/\mathbf{Tr}_{\mathrm{A}}\left(\rho_{\mathrm{A}}^{q}\right)\right]^{2}\right\}$. When q gets close to 1, the difference between the two curves is reduced. For system B, the curves are similar.

In this chapter, we have analyzed the calculation of the quantum mean values for a two spin-½ system. The procedure can be generalized to an arbitrary compound system. Obviously, the targets are the spins of Mn^{3+} and Mn^{4+}, $S_A = 2$ and $S_B = 3/2$, respectively. We have also learned a new method (in accordance with the partial trace), which allows the above calculations in the full Hilbert space to be equivalent to the same calculations in the Hilbert subspaces.

PART III

APPLICATIONS

CHAPTER FOUR

APPLICATION OF THE MATRICES *E*: A MODEL FOR $Pr_{1-x}Ca_xMnO_3$ WITH $x = 0.30–0.50$

In the literature on manganites, there is a wide variety of phenomenological models, which seek to explain the many exotic features of these compounds. Although so far no model comprehends all the properties of these materials, several theories have emerged. For instance, we have the models already explained in Chapter One: double exchange, polaron effect, phase separation in a percolative process, and Griffiths singularity. However, several aspects of manganites constitute unsolved problems. In this sense, nonextensive statistical mechanics emerges as an alternative for studying them; this theory has proven to be a guide to investigating them.

Evidence for Nonextensive Magnetism in Manganites

Before using nonextensive statistical mechanics, there are some essential requirements that a system must satisfy: fractal geometry, long-range interaction or memory, inhomogeneities, and dissipation. There is evidence that manganites have these ingredients. Just to cite one example, the Japanese Atsuo Satou and Masanori Yamanaka (2001) showed that manganites have magnetic clusters with fractal structure, namely, the Cantor set.

Furthermore, the Brazilian researcher Mario de Souza Reis Jr. (2003) provided evidence that manganites are nonextensive objects: by using a spin-½ system subjected to a uniform magnetic field, Brillouin generalized functions can be obtained. Thus, the experimental behavior of $La_{2/3}Ca_{1/3}MnO_3$ and $Pr_{0.5}Ca_{0.5}Mn_{0.95}Ga_{0.05}O_3$ can be understood through nonextensive statistical mechanics. For the first compound, it has been shown that the anomaly appearing on the curve $1/M$ vs. T corresponds to nonextensive effects for $q = 0.1$ and spin 5/2; concerning the second compound, it has been shown that the curve M vs. H can be explained qualitatively via generalized Brillouin functions $B_{\text{sat.}}^{q}$ with q between 0.1 and 0.3 ('sat.' means saturation). Also, the $La_{0.60}Y_{0.07}Ca_{0.33}MnO_3$ manganite can be studied by using a classical model with generalized Langevin functions. As additional information, we point out that, outside the nonextensive context, there are experimental results and theoretical models regarding $Pr_{1-x}Ca_xMnO_3$ with x = 0.20–0.45; for example, it is possible to study the influence of the charge ordering in the magnetocaloric effect as well as address the tunnel effect in metal-manganites interfaces as $Au_{0.9}Pd_{0.1}/Pr_{1-x}Ca_xMnO_3$, $Al/Pr_{1-x}Ca_xMnO_3$, and silver-ink/$Pr_{1-x}Ca_xMnO_3$.

Finally, it must be indicated that, by using the $La_{0.7}Sr_{0.3}MnO_3$ data collected with tunneling microscopy, it has been possible to determine the entropic index q; as a result, this parameter represents the degree of magnetic inhomogeneity.

Experimental Curves

The $Pr_{0.65}Ca_{0.35}MnO_3$ measurements to be shown below have been taken with a commercial MPMS SQUID magnetometer, available at the Brazilian Center for Research in Physics (MPMS is the acronym for Magnetic Property Measurement System). Table 4.1 shows the values of mass, temperature, and external field used in the measurements indicated. In obtaining experimental data, the SQUID magnetometer, properly programmed, used a ZFC process, i.e., the SQUID, in an automated way, cooled the sample without the presence of an external magnetic field. Subsequently, the SQUID warmed the sample in the presence of an external magnetic field applied —the SQUID collects data concerning magnetic field strength h; specifically, the collected magnetic field is generated by a current produced by superconducting coils connected to the SQUID.

The experimental behavior of the series measured presents anomalies in the magnetization; for concentrations x = 0.30–0.50, an unusual drop is observed at low temperatures. Thereupon, in Figure 4.1, for $Pr_{1-x}Ca_xMnO_3$ with x = 0.35, we have magnetization vs. temperature; the temperature interval is from 10 to 300 K and the magnetic field interval, from 0 to 50 kOe (applied after cooling without a magnetic field). emu is the magnetic unit used by the SQUID magnetometer; after dividing emu by mass, we have emu/g (bear in mind 1 emu = 10^{-3} JT^{-1} = 1.0783 x 10^{20} μ_{Bohr}). Before introducing a model based on nonextensive statistical mechanics, next, we will see how this type of graph is analyzed in the scientific literature. For example, V. Hardy et al. (2001), who made similar measurements for x = 0.37, interpret that these compounds exhibit a metamagnetic transition under the application of weak magnetic fields. For x = 0.35, in the presence of magnetic fields ranging from 0 to 50 kOe, there are three peaks.

Table 4.1. Data used in the measurements with the SQUID magnetometer.

$Pr_{1-x}Ca_xMnO_3$			
Concentration x (adim.)	**Mass (g)**	**Temperature (K)**	**External field (kOe)**
0.35	0.0046	10–300	0–50
0.40	0.0033	10–300	0–50

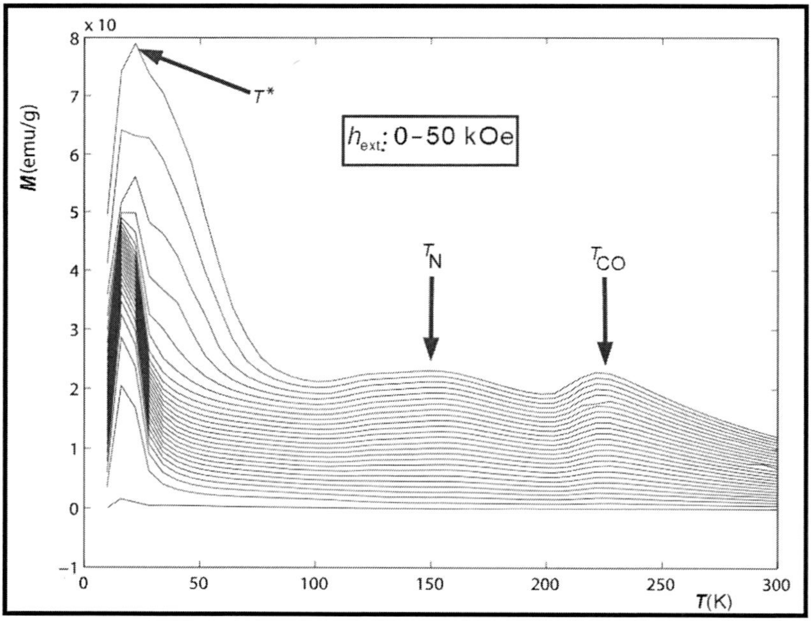

Figure 4.1. ZFC experimental curves *M* vs. *T* measured with the SQUID magnetometer for $Pr_{1-x}Ca_xMnO_3$ with x = 0.35. The signaled peaks correspond to three critical temperatures of three phase transitions, respectively.

The first peak, maximum, is around $T^* = 22$ K (T^* is different from the temperature defined in Eq. (2.55)). Below this temperature, we have a metastable metal-ferromagnetic phase, which turns out to be a first-order phase transition from an inhomogeneous phase, after the application of the external magnetic field. This last phase is composed of insulator ferromagnetic clusters embedded in an insulator antiferromagnetic matrix. Measurements with the FC procedure show that, in this case, the fall is much less pronounced (this difference between the two procedures is usually found in spin glasses, magnetization curves, and magnetic susceptibility curves). This fact suggests the existence of magnetic frustration at low temperatures.

The second peak is at $T_{\mathrm{N}} = 149$ K (Néel temperature). Below this value, we find a spin ordering, specifically, an antiferromagnetic structure; this antiferromagnetic arrangement is a CE-pseudotype.

The third peak is approximately at $T_{\mathrm{CO}} = 221$ K. Below this temperature but above T_{N}, there is a charge ordering (CO). Above T_{CO}, we have a paramagnetic insulator phase. When the applied magnetic field is low, on the order of 0.2–0.5 T, we observe the disappearance of the antiferromagnetic phase as well as the elimination of the charge-ordering phase.

Similarly, in Figure 4.2, we have magnetization versus external magnetic field, M vs. $h_{\mathrm{ext.}}$ ('ext.' means external). Below $T^* = 22$ K, we realize the initial angular coefficient is constant and determined by the demagnetizing factor; above this temperature, we observe a change in the angular coefficient. This is a signal of a phase transition, i.e., we move on from a ferromagnetic to an antiferromagnetic phase. Above $T_{\mathrm{N}} = 149$ K, we pass to a charge-ordering state; and surpassing $T_{\mathrm{CO}} = 221$ K, we observe a paramagnetic state.

Likewise, Figure 4.3 shows $\Delta S^{\mathrm{mag.}}_{\Delta h_{\mathrm{ext.}}}$ vs. T; $\Delta S^{\mathrm{mag.}}_{\Delta h_{\mathrm{ext.}}}$ is the magnetic entropy variation in the magnetic field variation $\Delta h_{\mathrm{ext.}} = 50$ kOe. The abbreviation 'mag.' means magnetic. We observe critical temperatures, which confirm the observations made in the graphics of magnetization. To calculate this magnetic entropy, we will use the following Maxwell relation for magnetic systems:

$$\left(\frac{\partial S^{\mathrm{mag.}}_{(T,h_{\mathrm{ext.}})}}{\partial h_{\mathrm{ext.}}}\right)_T = \left(\frac{\partial M_{(T,h_{\mathrm{ext.}})}}{\partial T}\right)_{h_{\mathrm{ext.}}}, \quad (4.1)$$

where $S^{\mathrm{mag.}}_{(T,h_{\mathrm{ext.}})}$ is the magnetic entropy at temperature T. Integrating both sides of this equation, we get the following expression:

$$\Delta S^{\mathrm{mag.}}_{\Delta h_{\mathrm{ext.}}} = \int_{h^i_{\mathrm{ext.}}}^{h^f_{\mathrm{ext.}}} dS^{\mathrm{mag.}}_{(T,h_{\mathrm{ext.}})_T} = \int_{h^i_{\mathrm{ext.}}}^{h^f_{\mathrm{ext.}}} \left(\frac{\partial M_{(T,h_{\mathrm{ext.}})}}{\partial T}\right)_{\Delta h_{\mathrm{ext.}}} dh_{\mathrm{ext.}}\,. \quad (4.2)$$

As the magnetization and the external magnetic field are experimental data, obtained in the SQUID experiment, we can numerically integrate the above equation by using the trapezoidal rule. The work of Vitalij K. Pecharsky and Karl A. Gschneidner Jr. (1999), in studying the magnetocaloric effect, provides details about that numerical integration.

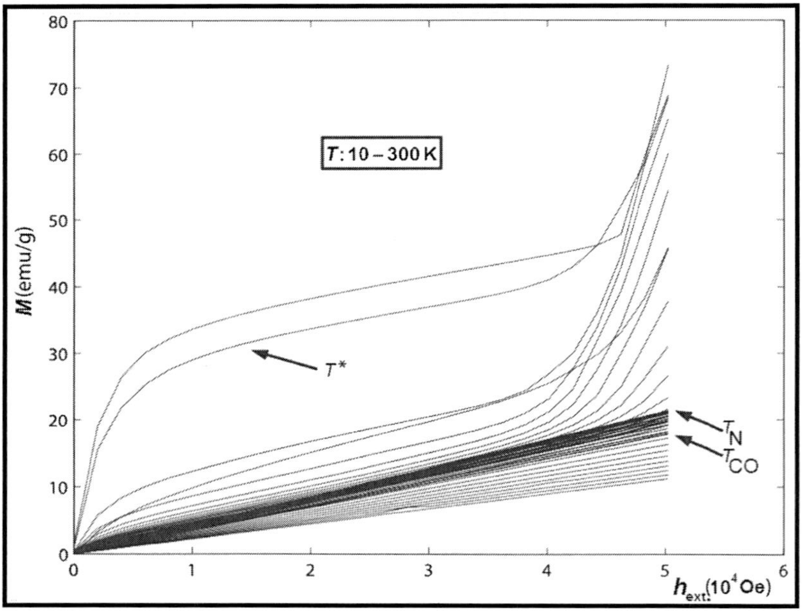

Figure 4.2. Experimental curves M vs. $h_{ext.}$ measured with the SQUID magnetometer for $Pr_{1-x}Ca_xMnO_3$ with x = 0.35. The signals indicate a change in the slope of the tangent line; these changes happen in the corresponding phase transition temperatures.

On the other hand, there is a behavior similar to the magnetization of the $Pr_{1-x}Ca_xMnO_3$ series, which was observed in the $La_{0.225}Ca_{0.375}Pr_{0.40}MnO_3$ manganites. Thus, magnetic phases have been identified through the temperature dependence of the magnetization by using $B_{ext.}$=1 tesla. Besides, the $Ca_{1-x}Sm_xMnO_3$ series, with x = 0.15, 0.16, and 0.2, presents a similar behavior, but with temperatures higher than those for the $Pr_{1-x}Ca_xMnO_3$ series.

In the perovskite cobaltites, cobalt oxides, in $La_{1-x}Sr_xCoO_3$ (with x = 0.2, 0.3, and 0.5, and $h_{ext.}$=20 Oe), it has been observed a behavior similar to the $Pr_{1-x}Ca_xMnO_3$ magnetization with x = 0.35–0.50. In the scientific literature, it is stated that, in addition to this similarity, there are many common features between manganites and cobaltites; both oxides possess very unusual magnetic and transport properties, which are determined by the competition among various interactions like double exchange, superexchange, and the Jahn-Teller effect (all already shown in Chapter One). The charge ordering, metal-insulator transition, high values of magnetoresistance, and spin-glass state are examples of such properties. Besides, in the $Al_{49}Fe_{30}Cu_{21}$ alloy with magnetic fields of 50 and 500 Oe, the irreversibility of the ZFC and FC magnetization has been found. This, as mentioned above, is a signal of a possible spin-glass behavior. Meanwhile, this feature, too, could be explained in the context of the theory of superparamagnetism.

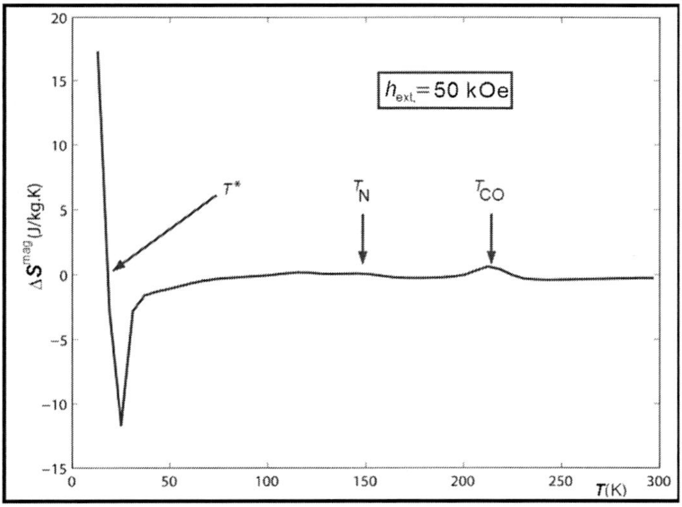

Figure 4.3. Experimental magnetic entropy, $\Delta S^{\mathrm{mag.}}$ vs. T, for $Pr_{1-x}Ca_xMnO_3$ with x = 0.35.

Quasi-probability Distributions

The relaxation of the Tsallis cutoff, Eq. (2.6), leads to probabilities with a value other than a positive real number; therefore, we will learn some precedents on this issue. The first quasi-probability distributions were proposed in 1932 by the Hungarian American theoretical physicist Eugene Paul Wigner, who would win the 1963 Nobel Prize in Physics. These distributions originated from the impossibility of representing the quantum states of a particle in the phase space, which is a consequence of the Heisenberg uncertainty principle. It is common knowledge that the Wigner distributions do not satisfy all the properties of a probability distribution and may take even negative values. In 1948, Jean Ville introduced this idea in signal analysis; their importance is now recognized since these distributions provide a powerful theoretical basis for time-frequency analysis; namely, their discrete-time feature constitutes an eminently practical tool for signal analysis.

Quasi-probability functions are widely used in the study of quantum systems. In this context, its main use is to analyze the connection between classical and quantum descriptions. Furthermore, it is known the proposal of the American theoretical physicist Richard Phillips Feynman to introduce negative probabilities to study the double-slit Young experiment.

Thus, we have shown the existence of antecedents in the use of negative probabilities. In our case, these ones appear when we relax the Tsallis cutoff.

Nonextensive Quasi-probability Distributions

Since manganites meet the requirements to use nonextensive statistical mechanics, we will now learn a phenomenological model stemmed from relaxing the Tsallis cutoff. This model can be applied to magnetization experimental curves considering weak magnetic fields, e.g., 2 and 4 kOe; see Figure 4.4. In these curves, we can observe the disappearance of the peaks indicating charge-ordered and antiferromagnetic phases, respectively.

Therefore, below, we describe the mathematical tool that will be used. The relaxation of the Tsallis cutoff leads to nonextensive quasi-probability distributions. By virtue of Eq. (2.6), the Tsallis cutoff is defined as the following condition:

$$[1-(1-q)\beta'\varepsilon_i]_+ = \begin{cases} 1-(1-q)\beta'\varepsilon_i, & \text{if } [1-(1-q)\beta'\varepsilon_i] > 0 \\ 0, & \text{else} \end{cases}; \quad (4.3)$$

we realize the relaxation of this condition implies that the contributions of $[1-(1-q)\beta'\varepsilon_i \leq 0]$ will be also considered. Consequently, we can define a density matrix of nonextensive quasi-probability E^{R} as a real matrix, where 'R' means real, namely,

$$E^{\mathrm{R}} = \mathbf{Re}\left\{\frac{[1-(1-q)\beta'\varepsilon_i]^{\frac{1}{1-q}}}{Z_q}\right\}, \quad (4.4)$$

the formulae for β' and Z_q are similar to Eqs. (2.4) and (2.5), respectively. However, now these parameters can have complex values. The traditional concept of populations maintains for $q = 1$ (Boltzmann-Gibbs) because the aforementioned cutoff is not considered in that case.

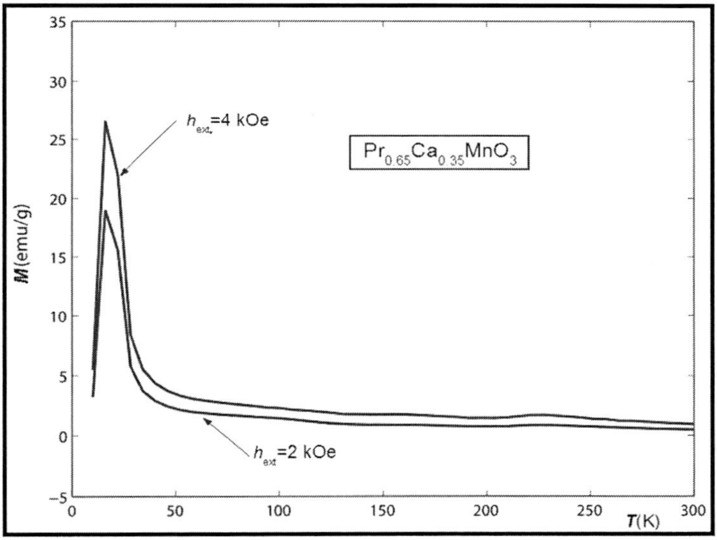

Figure 4.4. Experimental curves M vs. T for $Pr_{0.65}Ca_{0.35}MnO_3$ under low magnetic fields.

A Model for $Pr_{1-x}Ca_xMnO_3$ (x = 0.30–0.50) Under Weak Magnetic Fields

Preliminary Formulae

Before introducing the specific model, we will learn some formulae for a compound system A + B + C because our manganite compound has three constituents: Mn^{3+}, Mn^{4+}, and Pr^{3+} (O and Ca are irrelevant to our calculations). Consequently, the nonextensive magnetizations in the full Hilbert space are defined as follows:

$$M_{\mathrm{A}} = \frac{\mathbf{Tr}(g_{\mathrm{A}} E^{\mathrm{R}} J_z^{\mathrm{A}})}{\mathbf{Tr}(E^{\mathrm{R}})} \frac{\mu_{\mathrm{Bohr}}}{m_{\mathrm{A}}}, M_{\mathrm{B}} = \frac{\mathbf{Tr}(g_{\mathrm{B}} E^{\mathrm{R}} J_z^{\mathrm{B}})}{\mathbf{Tr}(E^{\mathrm{R}})} \frac{\mu_{\mathrm{Bohr}}}{m_{\mathrm{B}}}, \text{and } M_{\mathrm{C}} = \frac{\mathbf{Tr}(g_{\mathrm{C}} E^{\mathrm{R}} J_z^{\mathrm{C}})}{\mathbf{Tr}(E^{\mathrm{R}})} \frac{\mu_{\mathrm{Bohr}}}{m_{\mathrm{C}}}, \quad (4.5)$$

where g_{A}, g_{B}, and g_{C} are the Landé factors; μ_{Bohr}, the Bohr magneton; J_z^{A}, J_z^{B}, and J_z^{C}, spin matrices in the full Hilbert space. In the Hilbert subspaces, the nonextensive magnetizations are defined thus:

$$M_{\mathrm{A}} = \frac{\mathbf{Tr}_{\mathrm{A}}(g_{\mathrm{A}} E_{\mathrm{A}}^{\mathrm{R}} \mathrm{j}_z^{A})}{\mathbf{Tr}_{\mathrm{A}}(E_{\mathrm{A}}^{\mathrm{R}})} \frac{\mu_{\mathrm{Bohr}}}{m_{\mathrm{A}}}, M_{\mathrm{B}} = \frac{\mathbf{Tr}_{\mathrm{B}}(g_{\mathrm{B}} E_{\mathrm{B}}^{\mathrm{R}} \mathrm{j}_z^{B})}{\mathbf{Tr}_{\mathrm{B}}(E_{\mathrm{B}}^{\mathrm{R}})} \frac{\mu_{\mathrm{Bohr}}}{m_{B}}, \text{and } M_{\mathrm{C}} = \frac{\mathbf{Tr}_{\mathrm{C}}(g_{\mathrm{C}} E_{\mathrm{C}}^{\mathrm{R}} \mathrm{j}_z^{\mathrm{C}})}{\mathbf{Tr}_{\mathrm{C}}(E_{\mathrm{C}}^{\mathrm{R}})} \frac{\mu_{\mathrm{Bohr}}}{m_{\mathrm{C}}}, \quad (4.6)$$

$\mathrm{j}_z^{\mathrm{A}}$, $\mathrm{j}_z^{\mathrm{B}}$ and $\mathrm{j}_z^{\mathrm{C}}$ are the spin matrices in the corresponding Hilbert subspaces; $E_{\mathrm{A}}^{\mathrm{R}}$, $E_{\mathrm{B}}^{\mathrm{R}}$, and $E_{\mathrm{C}}^{\mathrm{R}}$ are the partial matrices of the nonextensive quasi-probability density matrix E^{R}:

$$E_{\mathrm{A}}^{\mathrm{R}} = \mathbf{Tr}_{(\mathrm{B,C})}(E^{\mathrm{R}}), \quad E_{\mathrm{B}}^{\mathrm{R}} = \mathbf{Tr}_{(\mathrm{A,C})}(E^{\mathrm{R}}), \text{ and } E_{\mathrm{C}}^{\mathrm{R}} = \mathbf{Tr}_{(\mathrm{A,B})}(E^{\mathrm{R}}). \quad (4.7)$$

Besides, the last-mentioned spin matrices are related to those from the full Hilbert space through

$$J_z^{\mathrm{A}} = \mathrm{j}_z^{\mathrm{A}} \otimes \mathbb{1}_{\mathrm{B}} \otimes \mathbb{1}_{\mathrm{C}}, \; J_z^{\mathrm{B}} = \mathbb{1}_{\mathrm{A}} \otimes \mathrm{j}_z^{\mathrm{B}} \otimes \mathbb{1}_{\mathrm{C}}, \text{and } J_z^{\mathrm{C}} = \mathbb{1}_{\mathrm{A}} \otimes \mathbb{1}_{\mathrm{B}} \otimes \mathrm{j}_z^{\mathrm{C}}, \quad (4.8)$$

where $\mathbb{1}_A$, $\mathbb{1}_B$, and $\mathbb{1}_C$ are the identity matrices in the corresponding Hilbert subspaces.

Analogously to the magnetizations, the nonextensive internal energies are defined in the full Hilbert space as follows:

$$U_{\mathrm{A}} = \frac{\mathbf{Tr}(E^{\mathrm{R}} H_{\mathrm{A}})}{\mathbf{Tr}(E^{\mathrm{R}})}, \quad U_{\mathrm{B}} = \frac{\mathbf{Tr}(E^{\mathrm{R}} H_{\mathrm{B}})}{\mathbf{Tr}(E^{\mathrm{R}})}, \text{ and } \quad U_{\mathrm{C}} = \frac{\mathbf{Tr}(E^{\mathrm{R}} H_{\mathrm{C}})}{\mathbf{Tr}(E^{\mathrm{R}})}, \quad (4.9)$$

where H_{A}, H_{B}, and H_{C} stand for the Hamiltonian matrices, for the corresponding subsystems, in the full Hilbert space. In the Hilbert subspaces, the internal energies are defined this way:

$$U_{\mathrm{A}} = \frac{\mathbf{Tr}_{\mathrm{A}}(E_{\mathrm{A}}^{\mathrm{R}} \mathcal{H}_{A})}{\mathbf{Tr}_{\mathrm{A}}(E_{\mathrm{A}}^{\mathrm{R}})}, \quad U_{B} = \frac{\mathbf{Tr}_{\mathrm{B}}(E_{\mathrm{B}}^{\mathrm{R}} \mathcal{H}_{B})}{\mathbf{Tr}_{\mathrm{B}}(E_{\mathrm{B}}^{\mathrm{R}})}, \text{and } U_{C} = \frac{\mathbf{Tr}_{\mathrm{C}}(E_{\mathrm{C}}^{\mathrm{R}} \mathcal{H}_{\mathrm{C}})}{\mathbf{Tr}_{\mathrm{C}}(E_{\mathrm{C}}^{\mathrm{R}})}, \quad (4.10)$$

where $\mathcal{H}_{\mathrm{A}}$, $\mathcal{H}_{\mathrm{B}}$, and $\mathcal{H}_{\mathrm{C}}$ are the Hamiltonian matrices in the Hilbert subspaces, respectively. These matrices are related to the ones in the full Hilbert space through the following expressions:

$$H_{\mathrm{A}} = \mathcal{H}_{\mathrm{A}} \otimes \mathbb{1}_{\mathrm{B}} \otimes \mathbb{1}_{\mathrm{C}}, H_{\mathrm{B}} = \mathbb{1}_{\mathrm{A}} \otimes \mathcal{H}_{\mathrm{B}} \otimes \mathbb{1}_{\mathrm{C}}, \text{and } H_{\mathrm{C}} = \mathbb{1}_{\mathrm{A}} \otimes \mathbb{1}_{\mathrm{B}} \otimes \mathcal{H}_{\mathrm{C}}. \quad (4.11)$$

The Model

The previously outlined theoretical framework can be applied to study manganites because nonextensive statistical mechanics is useful in understanding the manganese oxides. Thereupon, in our model, we will consider both the Oguchi model in the mean-field approximation and the contribution of the crystal field due to praseodymium.

The Oguchi model takes into account two spins interacting exactly, plus the interaction of these two spins with an average magnetic field; the mean-field Heisenberg model, used in Chapter Three, considers that each spin is independent but interacting with an effective field. Concerning the crystal field, originated from electrostatic interactions in a crystal, in the 3d ions, it provokes the lifting of the degeneracy of the $\mathrm{t_{2g}}$ and $\mathrm{e_g}$ orbitals. This was shown illustratively in Figure 1.2 in Chapter One. Then, as a guide to the study of the $Pr_{1-x}Ca_xMnO_3$ manganites, we will use a crystal field of cubic symmetry; models more sophisticated can consider the splitting of a crystal field with tetragonal symmetry. For the magnetization, we take into account the contributions of magnetic manganese and praseodymium (specifically, manganese with valence states Mn^{3+} and Mn^{4+} having concentrations x and 1-x, respectively, and praseodymium with valence state Pr^{3+}). Thus, we have the following Hamiltonian:

$$H_{\text{mag.}} = H_{\text{Ogu.}} + H_{\text{cf}}, \qquad (4.12)$$

where $H_{\text{Ogu.}}$ is the Oguchi term and H_{cf}, the contribution of the crystal field.

Concerning the former, it is given by

$$H_{\text{Ogu.}} = -2J\vec{S}_{\text{Mn}^{3+}}.\vec{S}_{\text{Mn}^{4+}} - \mu_{\text{Bohr}}(S^z_{\text{Mn}^{3+}} + S^z_{\text{Mn}^{4+}})(B_{\text{Mn}^{3+}} + B_{\text{Mn}^{4+}}), \qquad (4.13)$$

$B_{\text{Mn}^{3+}}$ is the field acting on the Mn^{3+} ion:

$$B_{\text{Mn}^{3+}} = B_0 + \lambda_{\text{Mn}^{3+}}\text{M}_{\text{Mn}^{3+}} + \lambda_{\text{Mn}^{3}-\text{Mn}^{4+}}\text{M}_{\text{Mn}^{4+}} \qquad (4.14),$$

equally for Mn^{4+}

$$B_{\text{Mn}^{4+}} = B_0 + \lambda_{\text{Mn}^{4+}}\text{M}_{\text{Mn}^{4+}} + \lambda_{\text{Mn}^{3}-\text{Mn}^{4+}}\text{M}_{\text{Mn}^{3+}}, \qquad (4.15)$$

where $\lambda_{\text{Mn}^{3+}}$ and $\lambda_{\text{Mn}^{4+}}$ represent the intra-lattice coupling, i.e., interactions inside each manganese sublattice; $\lambda_{\text{Mn}^{3}-\text{Mn}^{4+}}$ symbolizes the interlattice coupling, i.e., interactions between the corresponding sublattices of Mn^{3+} and Mn^{4+}. Besides, ($\lambda_{\text{Mn}^{3}-\text{Mn}^{4+}} > 0$) means a ferromagnetic coupling and ($\lambda_{\text{Mn}^{3}-\text{Mn}^{4+}} < 0$), an antiferromagnetic coupling.

Concerning the latter, it is the contribution of the crystal field with cubic symmetry from the Pr^{3+} ion. The form of H_{cf} depends on the choice of the axes. If we consider that the Z-axis coincides with the crystal direction (100) (Miller indices), we have the following expression:

$$H_{\text{cf}} = W\left\{\frac{X}{F_4}(O_4^0 + 5O_4^4) + \left(\frac{1-|X|}{F_6}\right)(O_6^0 + 21O_6^4)\right\}, \qquad (4.16)$$

where O_4^0, O_4^4, O_6^0, and O_6^4 are the Stevens operators for a cubic symmetry, which are shown in Table 4.2. These operators are utilized to quantify the effect of the crystal field on magnetization. Specifically, quenching appears due to the previously mentioned field, i.e., the attenuation of both the angular momentum and the magnetization happens. Thereupon, W, X, F_4, and F_6 are parameters having the following relationships:

$$B_4^0 F_4 = WX \quad \text{and} \quad B_6^0 F_6 = W(1-|X|) \qquad (4.17)$$

or

$$\frac{X}{1-|X|} = \frac{F_4 B_4^0}{F_6\, B_6^0}, \qquad (4.18)$$

where $-1 < X < 1$; B_4^0 and B_6^0 are additional parameters participating in the following alternative notation for Eq. (4.16):

$$H_{\text{cf}} = B_4^0(O_4^0 + 5O_4^4) + B_6^0(O_6^0 + 21O_6^4); \qquad (4.19)$$

however, the calculation of the factors B_4^0 and B_6^0 involves a high degree of imprecision (for these parameters contain terms with $\langle r^n \rangle$, i.e. n-th order ionic radius, and electron shielding factors, too). In the scientific literature, relationship (4.18) is very used because the $\frac{B_4^0}{B_6^0}$ ratio has no similar difficulty. The calculation by Lea, Leask, and Wolf (1962) is a good reference to the optimal values of these parameters. For Pr^{3+}, with total angular momentum quantum number $J = 4$, the above reference indicates $F_4 = 60$ and $F_6 = 1260$. Then, because Pr^{3+} has the quantum number $J = 4$ (with orbital angular momentum quantum number $L = 5$ and spin angular momentum quantum number $S = 1$), the compound system $Mn^{3+} + Mn^{4+} + Pr^{3+}$ must be represented in the 180x180 full Hilbert space, as shown in Table 4.3.

Table 4.2. For a cubic symmetry, the conventional Stevens operators. J is the total angular momentum quantum number; J_x, J_y, and J_z, components of the total angular momentum operator; $J_\pm$, creation and annihilation operators, respectively.

Stevens Operators
$O_4^0 = 35J_z^4 - (30X - 25)J_z^2 + 3X^2 - 6X$
$O_4^4 = \frac{1}{2}(J_+^4 + J_-^4)$
$O_6^0 = 231J_z^6 - (315X - 735)J_z^4 + (105X^2 - 525X + 294)J_z^2 - 5X^3 + 40X^2 - 60X$
$O_6^4 = \frac{1}{4}[11J_z^2 - X - 38](J_+^4 + J_-^4) + (J_+^4 + J_-^4)[11J_z^2 - X - 38]$
* $X \equiv J(J + 1)$ and $J_\pm \equiv J_x \pm iJ_y$

Table 4.3. Angular momentum quantum numbers of $Pr_{0.65}Ca_{0.35}MnO_3$. Manganese has mixed valence. HS means Hilbert subspace and FHS, full Hilbert space.

	$Pr_{1-x}Ca_xMnO_3$	
Mn^{3+}	**Mn^{4+}**	**Pr^{3+}**
$S = 2$	$S = 3/2$	$J = 4$
$2S + 1 = 5$	$2S + 1 = 4$	$2J + 1 = 9$
HS: 5x5	HS: 4x4	HS: 9x9
	FHS: 180x180	

To simplify the calculations, these ones were performed by a computer program, where was used a hypothetical $J_{Pr^{3+}} = 3/2$, producing the

80x80 full Hilbert space. The objective of the model introduced is to show how the relaxation of the Tsallis cutoff leads to a drop in the magnetization at low temperatures.

After using nonextensive quasi-probability matrices E^R, in Figure 4.5, we contrast the experimental and computational results for $Pr_{0.65}Ca_{0.35}MnO_3$; in both curves, the magnetic field is $H_{ext.}$= 2 kOe. Furthermore, for $Pr_{0.7}Ca_{0.3}MnO_3$, with the same theoretical model used for Figure 4.5, in Figure 4.6, we have magnetization versus temperature but with external magnetic field $H_{ext.}$= 4 kOe. This last graph shows, once more, that the relaxation of the Tsallis cutoff provokes the fall of the magnetization at low temperatures. The model of the matrices E^R indicates that nonextensive statistical mechanics is useful to study manganites, which is significant because the understanding of manganites will allow a better explanation of the superconducting perovskites, which have a high critical temperature.

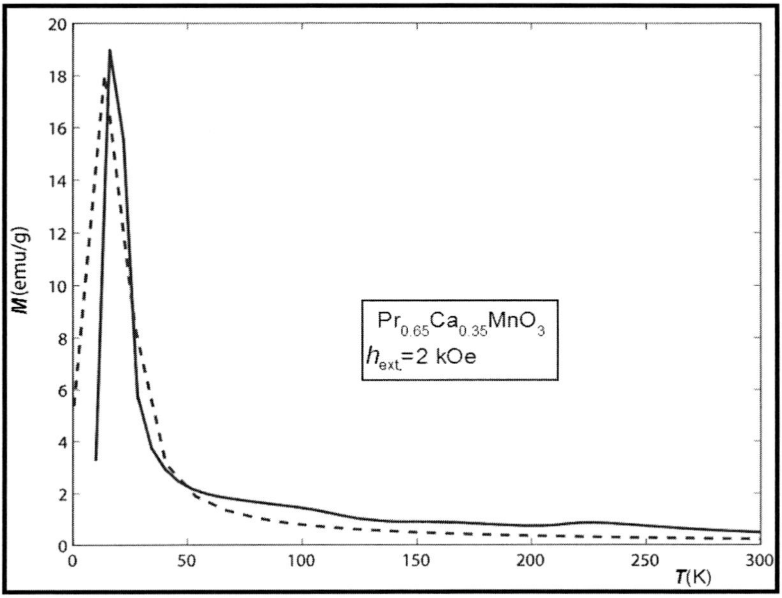

Figure 4.5. For $Pr_{0.65}Ca_{0.35}MnO_3$, a comparison between experimental and calculated M vs. T. The solid line represents the experimental curve; the dashed line, the computer calculation. This last one was made with matrices E^R for q = 0.55.

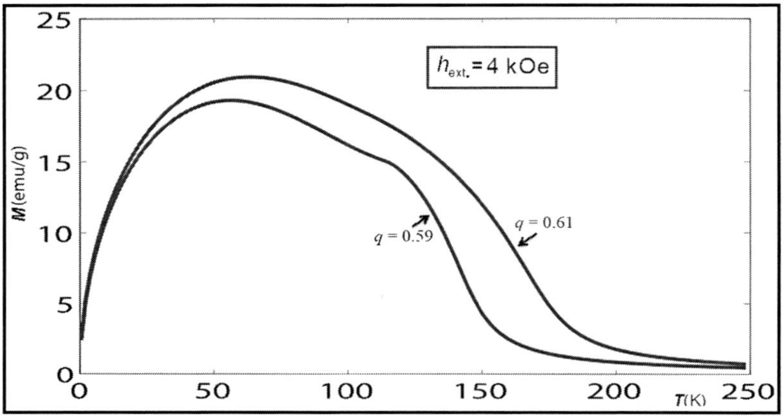

Figure 4.6. For $Pr_{0.70}Ca_{0.30}MnO_3$, computer calculation of M vs. T with matrices E^{R}; the values of q are indicated. These curves are similar to the experimental curves reported in the literature (for example, see Deac, Mitchell, and Schiffer 2001).

This chapter showed a phenomenological model that explains the magnetization anomaly in $Pr_{1-x}Ca_xMnO_3$ with x = 0.30–0.50, in the presence of low magnetic fields. The model uses nonextensive quasi-probabilities, which provoke the emergence of a ferromagnetic, irreversible metastable phase at low temperatures with concentration x = 0.30–0.50. The qualitative agreement between the computer calculations and the experimental curves, along with many clues existing in the scientific literature, indicate that manganites present magnetic nonextensiveness.

CHAPTER FIVE

IN SUMMARY

In this book, we have dealt with the $Pr_{1-x}Ca_xMnO_3$ manganites. Nonextensive statistical mechanics is suitable for a description of these materials; we have learned two interesting contributions to the research of these compounds.

On the one hand, we learned a method different from the standard to calculate mean values of compound quantum systems: the method of the matrices E. For instance, with this method, in the third version of nonextensive statistical mechanics, the Hilbert subspace calculations for a system A + B agree with the full Hilbert space calculations for the same compound system. This alternative method is different from the procedure that would be natural in the third version of nonextensive statistical mechanics. A summary of the formal derivation from the full Hilbert space to the corresponding Hilbert subspace, as well as analytical calculations for specific cases, undoubtedly demonstrated that the matrices E are deciding to find the quantum mean values when compound systems are studied. Besides, by using these matrices, we showed that the escort distributions can only be used in the full Hilbert space, but not in the Hilbert subspaces.

On the other hand, we also knew a phenomenological model, which qualitatively explains the behavior of the magnetization in $Pr_{1-x}Ca_xMnO_3$, with concentrations $x = 0.30–0.50$. In this model, the relaxation of the Tsallis cutoff and the consequent nonextensive quasi-probability distributions are important. Thus, we have shown that nonextensive statistical mechanics is a guide to studying manganites; in that sense, the aforementioned model reinforces the evidence of a nonextensive behavior in manganites. The agreement between the computer calculations and the experimental curves is good.

Furthermore, the increasing quantity of papers regarding nonextensive statistical mechanics indicates that this theory is useful in explaining many anomalous phenomena.

APPENDIX A

TOPICS ON MAGNETISM

Magnetoresistance

Next, for illustrative purposes, we will describe some characteristics of the phenomenon of magnetoresistance.

Regular Magnetoresistance

This one is the change in the electrical resistance of a material when an external magnetic field is applied. The most basic example is the transverse magnetoresistance; it is associated with the classic Hall effect discovered, in 1879, by the American physicist Edwin Herbert Hall.

Concerning the last effect, in a conductor, the simultaneous applications of an electric field and a magnetic field, in the respective x- and z- directions, provoke the emergence of another electric field in the y-direction. This is due to the Lorentz force, which acts on the electrons moving along the x-axis. The new electric field provokes the Hall voltage. E. Hall carried out his experiments at room temperature with the application of magnetic fields below 1 tesla. The measurement of the Hall coefficient, R_H, is usually used to map the Fermi surface. Didactically, Kaganov & Lifshits (1979) expressed that this surface is "the stage on which the 'drama of the life of the electron' is played out."

On the other hand, in 1980, Klaus von Klitzing—in studying materials with highly mobile electrons along a surface, at very low temperatures and very powerful magnetic fields (~ 30 teslas)—discovered the quantum Hall effect, wherein the Hall resistance does not vary linearly, but in jumps, according to the applied magnetic field. The resistance values do not depend on the material but on fundamental physical constants divided by an integer; the material becomes a kind of superconductor. So, a new international resistance pattern emerged: 1 klitzing, defined as the Hall frequency in the fourth jump $\left(\frac{he^2}{4}\right)$. Furthermore, there is an anomalous Hall effect found in magnetic materials; the voltage is proportional to the magnetization. This effect is useful, for example, when measuring magnetic hysteresis.

Finally, in March 2004, the discovery of the spin-Hall effect in semiconductors was announced; nowadays, the dubbed spin Hall magnetoresistance has the interest of the investigators.

Giant Magnetoresistance

In 1988, the Argentine Brazilian physicist Mario Norberto Baibich and his collaborators discovered the giant magnetoresistance effect (GMR). Their work entitled *Giant magnetoresistance of (001)Fe/(001)Cr magnetic superlattices* was published in *Physical Review Letters*. M. Baibich et al. used the (001)Fe/(001)Cr super-lattice prepared by molecular beam epitaxy. The magnetoresistance measured was decreased by a factor of 2. Likewise, in 1989, the group of the German G. Binasch was the first to confirm the important discovery by publishing their article entitled *Enhanced magnetoresistance in layered magnetic structures with antiferromagnetic interlayer exchange.* Curiously, the French physicist Albert Fert and the German physicist Peter Grünberg won the 2007 Nobel Prize in Physics by discovering the giant magnetoresistance effect. A. Fert was coauthor along with M. N. Baibich in the aforementioned work, and Peter Grünberg was also coauthor along with G. Binash in the previously mentioned article.

Coincidentally, the two groups used Fe and Cr layers. Specifically, they used structures formed from Fe sandwiches filled with a layer of three Cr atoms. They also applied different magnetic fields, and, afterward, they measured electrical resistance values. When the outside-the-sandwich layers have magnetic alignment opposite each other, the structure has high electrical resistance. If the alignment is parallel, the resistance will be lower, approximately half of the previous configuration. A giant magnetoresistance experiment is analogous to a polarization experiment since the first magnetic layer allows the easy passage of electron spin state only if the second magnetic layer is aligned. If so, the spin channel easily goes through the structure, and the resistance lowers. However, if the second magnetic layer is not aligned, the spin channel does not traverse the structure and the resistance augments.

On the other hand, in addition to the giant magnetoresistance detected in multilayers, there is a giant magnetoresistance called granular. It occurs in granular solids, which are artificial systems where the grains are formed by magnetic elements (Fe, Ni, or Co) and embedded in a nonmagnetic metallic matrix (Cu, Ag, or Au). Giant magnetoresistance is usually observed in Cu matrices containing Co grains (these elements are known to be immiscible, i.e., they cannot form metal alloys). Furthermore, the granular solids do not reach the values of the strong change experienced by the giant magnetoresistance multilayers. In 1994, the giant magnetoresistance began to be applied in the automotive industry as part of a sensor for brake systems.

Today, the giant magnetoresistance is used in most of the read heads in hard disks for computers. Spintronics, or spin electronics, has been developed from the discovery of the giant magnetoresistance effect. Thereupon, spin valves are structures in which a layer is mobilized in a field, whereas another is not. There are several schemes of spin valves, mainly based on the phenomenon of exchange bias, which consists in the hysteresis loop displacement occurring at the interface between ferromagnetic and antiferromagnetic films. This displacement would be provoked by the exchange interaction in that interface. Films exhibiting this phenomenon are interesting because the displacement may be important to control the magnetization in technological instruments.

Magnetocaloric Effect

It is an important physical phenomenon discovered in iron by the German Physicist Emil Gabriel Warburg in 1881. It is a magneto-thermodynamic effect that we can find in solid materials; namely, in adiabatic conditions, when we apply or remove an external magnetic field from a solid, this presents an increase or decrease in its temperature, respectively. We must point out that, usually, the magnetocaloric effect happens when the material is near to its Curie temperature. A good classical review is the work of Gschneidner Jr. and Pecharsky (2000). A more recent review article is the work of A. M. Tishin et al. (2016).

To understand the magnetocaloric effect, we must consider that, in a paramagnetic material, the molecules have intrinsic magnetic dipole moments. Thereupon, on the one hand, when we apply a strong external magnetic field, the internal magnetic dipoles align according to the direction of the applied magnetic field. Thus, the material has a higher ordering, but its magnetic entropy has decreased. However, on the other hand, when we dealt with an adiabatic process, the material has increased its magnetic entropy. This is escorted by the movement of the atoms and, consequently, an increase in the temperature of the material, that is, the magnetocaloric effect has occurred.

Applications of the Magnetocaloric Effect

The great interest in magnetocaloric materials is because they have potential technological applications, above all in magnetic refrigeration with low environmental impact and high energy efficacy. In contrast to a traditional refrigerator, which is based on compressing and expanding a type of gas, a magnetic refrigerator is based on magnetizing and demagnetizing the magnetocaloric material. Besides, it is common knowledge that prototypes of magnetic refrigerators have been developed since the 1930s. However, so far there are no household refrigerators because, in the prototypes to create the magnetocaloric effect, it is necessary to have superconducting magnets cooled to ultra-low temperatures.

Nonetheless, these promising materials also have applications in the execution of thermodynamic cycles and energy harvesting. This last application, also known as energy scavenging, consists in feeding a small device without being connected to the electric power. Even more, nowadays, there are many medical applications. For example, in the cancer treatment through the magnetic hyperthermia.

Concerning the specific material used in several applications, we have ferromagnetic perovskites (manganites), glass ceramics, gadolinium, and spinel ferrites. However, in the scientific literature, it has also been reported the use of compounds such as Fe-Rh, Gd-Si-Ge, La-Fe-Si, Ni-Mn-Ga, and Mn-As. Likewise, the use of the Heussler alloys has been reported; these ones are ternary intermetallic compounds having interesting properties. They have applications such as topological insulators and half-metallic ferromagnetic, etc. The family of the Heussler alloys consists of around a thousand compounds.

APPENDIX B

DEMONSTRATION OF SEVERAL PROPERTIES OF THE Q-LOGARITHM FUNCTION

Finding Out a Formula for the *q*-Logarithm of the Product of Two Functions

We will search a formula for $\mathrm{Ln}_q(y_1 y_2)$. Then, we start by considering the definition of the q-logarithm function:

$$\mathrm{Ln}_q(y_1 y_2) = \frac{(y_1 y_2)^{1-q} - 1}{1-q}, \qquad \text{(B.1)}$$

after using the power of a product property, it follows that

$$\mathrm{Ln}_q(y_1 y_2) = \frac{{y_1}^{1-q}\, {y_2}^{1-q} - 1}{1-q}\ ; \qquad \text{(B.2)}$$

now, because of the definition of the q-logarithm function, we know that

$${y_1}^{1-q} = (1-q)\mathrm{Ln}_q(y_1) + 1 \ \text{ and } \ {y_2}^{1-q} = (1-q)\mathrm{Ln}_q(y_2) + 1, \text{(B.3)}$$

replacing this last set of formulae into Eq. (B.2), we get this way:

$$\mathrm{Ln}_q(y_1 y_2) = \frac{[(1-q)\mathrm{Ln}_q(y_1)+1]\,[(1-q)\mathrm{Ln}_q(y_2)+1] - 1}{1-q}, \qquad \text{(B.4)}$$

and multiplying the terms inside the brackets, we have the following expression:

$$\mathrm{Ln}_q(y_1 y_2) = \frac{(1-q)^2\mathrm{Ln}_q(y_1)\mathrm{Ln}_q(y_2) + (1-q)\mathrm{Ln}_q(y_1) + (1-q)\mathrm{Ln}_q(y_2) + 1 - 1}{1-q}, \text{(B.5)}$$

now, we cancel $1-1$ and cross cancel $(1 - q)$, too. Then,

$$\mathrm{Ln}_q(y_1 y_2) = (1-q)\mathrm{Ln}_q(y_1)\mathrm{Ln}_q(y_2) + \mathrm{Ln}_q(y_1) + \mathrm{Ln}_q(y_2)\,, \text{(B.6)}$$

after changing the order of the addends, we have

$$\mathrm{Ln}_q(y_1 y_2) = \mathrm{Ln}_q(y_1) + \mathrm{Ln}_q(y_2) + (1-q)\mathrm{Ln}_q(y_1)\mathrm{Ln}_q(y_2)\,, \text{(B.7)}$$

this is the pseudoadditive formula for the q-logarithm of two functions. Likewise, on the right side, we identify the previously introduced operation of the q-sum, Eq. (2.256) in Chapter Two. Therefore, we obtain similarity with standard mathematics, that is,

$$\mathrm{Ln}_q(y_1 y_2) = \mathrm{Ln}_q(y_1) \oplus_q \mathrm{Ln}_q(y_2)\,, \qquad \text{(B.8)}$$

where evidently

$$\mathrm{Ln}_q(y_1) \oplus_q \mathrm{Ln}_q(y_2) = \mathrm{Ln}_q(y_1) + \mathrm{Ln}_q(y_2) + (1-q)\mathrm{Ln}_q(y_1)\mathrm{Ln}_q(y_2). \ \text{(B.9)}$$

On the other hand, for $q = 1$, we have as follows:

$$\mathrm{Ln}_1(y_1y_2) = \mathrm{Ln}_1(y_1)\oplus_1\mathrm{Ln}_1(y_2)$$
$$= \mathrm{Ln}_1(y_1) + \mathrm{Ln}_1(y_2) + (1 - 1)\mathrm{Ln}_1(y_1)\mathrm{Ln}_1(y_2)\,, \quad \text{(B.10)}$$

and simplifying, we obtain this way:

$$\mathrm{Ln}_1(y_1y_2) = \mathrm{Ln}_1(y_1)\oplus_1\mathrm{Ln}_1(y_2) = \mathrm{Ln}_1 y_1 + \mathrm{Ln}_1 y_2\,, \quad \text{(B.11)}$$

or more clearly,

$$\mathrm{Ln}(y_1y_2) = \mathrm{Ln}y_1 + \mathrm{Ln}y_2\,, \quad \text{(B.12)}$$

which is the expected standard result.

Finding Out a Formula for the q-Logarithm of the Product of Three Functions

Additionally, we now will search a formula for $\mathrm{Ln}_q(y_1y_2y_3)$, that is, using three variables. Thereupon, we will use two procedures for the same purpose.

On the one hand, concerning the first procedure, we have that

$$\mathrm{Ln}_q(y_1y_2y_3) = \mathrm{Ln}_q(ab)\,, \quad \text{(B.13)}$$

where we have set $a = y_1y_2$ and $b = y_3$. Thus, it follows that

$$\mathrm{Ln}_q(y_1y_2y_3) = \mathrm{Ln}_q(ab) = \mathrm{Ln}_q(a) + \mathrm{Ln}_q(b) + (1 - q)\mathrm{Ln}_q(a)\mathrm{Ln}_q(b), \quad \text{(B.14)}$$

after replacing the values of a and b,

$$\mathrm{Ln}_q(y_1y_2y_3) = \mathrm{Ln}_q(y_1y_2) + \mathrm{Ln}_q(y_3) + (1 - q)\mathrm{Ln}_q(y_1y_2)\mathrm{Ln}_q(y_3), \quad \text{(B.15)}$$

by applying the pseudoadditive formula for $\mathrm{Ln}_q(y_1y_2)$ again, it becomes thus:

$$\mathrm{Ln}_q(y_1y_2y_3) = \left[\mathrm{Ln}_q(y_1) + \mathrm{Ln}_q(y_2) + (1-q)\mathrm{Ln}_q(y_1)\mathrm{Ln}_q(y_2)\right] + \mathrm{Ln}_q(y_3)$$
$$+ (1 - q)\left[\mathrm{Ln}_q(y_1) + \mathrm{Ln}_q(y_2) + (1-q)\mathrm{Ln}_q(y_1)\mathrm{Ln}_q(y_2)\right]\mathrm{Ln}_q(y_3)\,, \quad \text{(B.16)}$$

or

$$\mathrm{Ln}_q(y_1y_2y_3) = \mathrm{Ln}_q(y_1) + \mathrm{Ln}_q(y_2) + (1 - q)\mathrm{Ln}_q(y_1)\mathrm{Ln}_q(y_2) + \mathrm{Ln}_q(y_3)$$
$$+(1 - q)\left[\mathrm{Ln}_q(y_1)\mathrm{Ln}_q(y_3) + \mathrm{Ln}_q(y_2)\mathrm{Ln}_q(y_3) + (1 - q)\mathrm{Ln}_q(y_1)\mathrm{Ln}_q(y_2)\mathrm{Ln}_q(y_3)\right] \text{(B.17)}$$

now, we will group terms. Then,

$$\mathrm{Ln}_q(y_1y_2y_3) = \mathrm{Ln}_q(y_1) + \mathrm{Ln}_q(y_2) + \mathrm{Ln}_q(y_3) + (1 - q)\mathrm{Ln}_q(y_1)\mathrm{Ln}_q(y_2)$$
$$+ (1 - q)\mathrm{Ln}_q(y_1)\mathrm{Ln}_q(y_3) + (1 - q)\mathrm{Ln}_q(y_2)\mathrm{Ln}_q(y_3) + (1 - q)^2\mathrm{Ln}_q(y_1)\mathrm{Ln}_q(y_2)\mathrm{Ln}_q(y_3); \text{(B.18)}$$

thus, the q-logarithm of the product of three functions is defined as the q-sum of the three corresponding logarithms, that is,

$$\mathrm{Ln}_q(y_1y_2y_3) = \mathrm{Ln}_q(y_1)\oplus_q\mathrm{Ln}_q(y_2)\oplus_q\mathrm{Ln}_q(y_3)\,, \quad \text{(B.19)}$$

On the other hand, concerning the second procedure, we have the following expressions:

$$\mathrm{Ln}_q(y_1y_2y_3) = \frac{(y_1y_2y_3)^{1-q} - 1}{1-q} = \frac{{y_1}^{1-q}\,{y_2}^{1-q}\,{y_3}^{1-q} - 1}{1-q}\,; \quad \text{(B.20)}$$

besides, we already know that

$${y_1}^{1-q} = (1 - q)\mathrm{Ln}_q(y_1) + 1, \quad {y_2}^{1-q} = (1 - q)\mathrm{Ln}_q(y_2) + 1\,,$$

and $y_3^{1-q} = (1-q)\mathrm{Ln}_q(y_3) + 1$, (B.21)

replacing these three equations into the penultimate equation, we get

$$\mathrm{Ln}_q(y_1y_2y_3) = \frac{[(1-q)\mathrm{Ln}_q(y_1)+1]\,[(1-q)\mathrm{Ln}_q(y_2)+1][(1-q)\mathrm{Ln}_q(y_3)+1]\ -1}{1-q}; \quad \text{(B.22)}$$

thereupon, obviously, it must be verified that

$$\mathrm{Ln}_q(y_1)\oplus_q\mathrm{Ln}_q(y_2)\oplus_q\mathrm{Ln}_q(y_3) = $$
$$= \frac{[(1-q)\mathrm{Ln}_q(y_1)+1]\,[(1-q)\mathrm{Ln}_q(y_2)+1][(1-q)\mathrm{Ln}_q(y_3)+1]\ -1}{1-q}. \quad \text{(B.23)}$$

On the other hand, for $q = 1$, we have the following expression:

$$\mathrm{Ln}_1(y_1y_2y_3) = \mathrm{Ln}_1(y_1)\oplus_1\mathrm{Ln}_1(y_2)\oplus_1\mathrm{Ln}_1(y_3)\,, \quad \text{(B.24)}$$

after considering $\oplus_1 = +$, it follows that

$$\mathrm{Ln}(y_1y_2y_3) = \mathrm{Ln}(y_1) + \mathrm{Ln}(y_2) + \mathrm{Ln}(y_3)\,, \quad \text{(B.25)}$$

which is the expected standard formula.

Demonstration of a Formula for the q-Logarithm of n Functions

Looking at the previous results, we propose that

$$\mathrm{Ln}_q(y_1y_2\ldots y_n) = \mathrm{Ln}_q(y_1)\oplus_q\mathrm{Ln}_q(y_2)\oplus_q\ldots\oplus_q\mathrm{Ln}_q(y_n) \quad \text{(B.26)}$$

and

$$\mathrm{Ln}_q(y_1)\oplus_q\mathrm{Ln}_q(y_2)\oplus_q\ldots\oplus_q\mathrm{Ln}_q(y_n)$$
$$= \frac{[(1-q)\mathrm{Ln}_q(y_1)+1]\,[(1-q)\mathrm{Ln}_q(y_2)+1]\ldots[(1-q)\mathrm{Ln}_q(y_n)+1]\ -1}{1-q}, \quad \text{(B.27)}$$

where $n \geq 2$ and $n \in \mathbb{N}$. In order to demonstrate these formulae, in the framework of the aforementioned second procedure, we will use the method of mathematical induction.

For $n = 2$, we have already proven that

$$\mathrm{Ln}_q(y_1y_2) = \mathrm{Ln}_q(y_1)\oplus_q\mathrm{Ln}_q(y_2)\,, \quad \text{(B.28)}$$

also, we already know that

$$\mathrm{Ln}_q(y_1)\oplus_q\mathrm{Ln}_q(y_2) = \frac{[(1-q)\mathrm{Ln}_q(y_1)+1]\,[(1-q)\mathrm{Ln}_q(y_2)+1]\ -1}{1-q}; \quad \text{(B.29)}$$

therefore, for $n = 2$, it has been proven.

For $n = h$, the formula is given as follows:

$$\mathrm{Ln}_q(y_1y_2\ldots y_h) = \mathrm{Ln}_q(y_1)\oplus_q\mathrm{Ln}_q(y_2)\oplus_q\ldots\oplus_q\mathrm{Ln}_q(y_h)\,, \quad \text{(B.30)}$$

and

$$\mathrm{Ln}_q(y_1)\oplus_q\mathrm{Ln}_q(y_2)\oplus_q\ldots\oplus_q\mathrm{Ln}_q(y_h)$$
$$= \frac{[(1-q)\mathrm{Ln}_q(y_1)+1][(1-q)\mathrm{Ln}_q(y_2)+1]\ldots[(1-q)\mathrm{Ln}_q(y_h)+1]\ -1}{1-q}, \quad \text{(B.31)}$$

we will assume that both formulae are true in accordance with the method of mathematical induction.

For $n = h + 1$, the two formulae are the following ones:

$$\mathrm{Ln}_q(y_1 \dots y_h y_{h+1}) = \mathrm{Ln}_q(y_1) \oplus_q \dots \oplus_q \mathrm{Ln}_q(y_h) \oplus_q \mathrm{Ln}_q(y_{h+1}), \quad \text{(B.32)}$$

and

$$\mathrm{Ln}_q(y_1) \oplus_q \dots \oplus_q \mathrm{Ln}_q(y_h) \oplus_q \mathrm{Ln}_q(y_{h+1}) = \\ = \frac{[(1-q)\mathrm{Ln}_q(y_1)+1] \dots [(1-q)\mathrm{Ln}_q(y_h)+1][(1-q)\mathrm{Ln}_q(y_{h+1})+1] - 1}{1-q}. \quad \text{(B.33)}$$

Thereupon, we start by proving formula (B.32). Hence, we will set the following parameters:

$$a = y_1 y_2 \dots y_h \quad \text{and} \quad b = y_{h+1}, \quad \text{(B.34)}$$

then, it follows that

$$\mathrm{Ln}_q(y_1 y_2 \dots y_h y_{h+1}) = \mathrm{Ln}_q(ab); \quad \text{(B.35)}$$

however, we already learned $\mathrm{Ln}_q(ab) = \mathrm{Ln}_q(a) \oplus_q \mathrm{Ln}_q(b)$, consequently

$$\mathrm{Ln}_q(y_1 y_2 \dots y_h y_{h+1}) = \mathrm{Ln}_q(ab) = \mathrm{Ln}_q(a) \oplus_q \mathrm{Ln}_q(b) \\ = [\mathrm{Ln}_q(y_1 y_2 \dots y_h)] \oplus_q \mathrm{Ln}_q(y_{h+1}), \quad \text{(B.36)}$$

now, we use the hypothesis inside the brackets; we obtain this way:

$$\mathrm{Ln}_q(y_1 y_2 \dots y_h y_{h+1}) = [\mathrm{Ln}_q(y_1) \oplus_q \mathrm{Ln}_q(y_2) \oplus_q \dots \oplus_q \mathrm{Ln}_q(y_h)] \oplus_q \mathrm{Ln}_q(y_{h+1}), \quad \text{(B.37)}$$

simplifying

$$\mathrm{Ln}_q(y_1 y_2 \dots y_h y_{h+1}) = \mathrm{Ln}_q(y_1) \oplus_q \mathrm{Ln}_q(y_2) \oplus_q \dots \oplus_q \mathrm{Ln}_q(y_h) \oplus_q \mathrm{Ln}_q(y_{h+1}), \quad \text{(B.38)}$$

which is the result we were looking for.

Likewise, to prove formula (B.33), we now will introduce two parameters:

$$r = \mathrm{Ln}_q(y_1) \oplus_q \dots \oplus_q \mathrm{Ln}_q(y_h) \quad \text{and} \quad s = \mathrm{Ln}_q(y_{h+1}), \quad \text{(B.39)}$$

consequently,

$$\mathrm{Ln}_q(y_1) \oplus_q \dots \oplus_q \mathrm{Ln}_q(y_h) \oplus_q \mathrm{Ln}_q(y_{h+1}) = r \oplus_q s = \frac{[(1-q)r+1][(1-q)s+1] - 1}{1-q}, \quad \text{(B.40)}$$

now, we bear in mind this second hypothesis:

$$\mathrm{Ln}_q(y_1) \oplus_q \dots \oplus_q \mathrm{Ln}_q(y_h) \\ = \frac{1}{1-q}\{[(1-q)\mathrm{Ln}_q(y_1) + 1][(1-q)\mathrm{Ln}_q(y_2) + 1] \dots [(1-q)\mathrm{Ln}_q(y_h) + 1] - 1\}, \quad \text{(B.41)}$$

Thus, we have

$$\mathrm{Ln}_q(y_1) \oplus_q \dots \oplus_q \mathrm{Ln}_q(y_h) \oplus_q \mathrm{Ln}_q(y_{h+1}) = r \oplus_q s \\ = \frac{1}{1-q}\left\{\left[(1-q)\frac{[(1-q)\mathrm{Ln}_q(y_1)+1][(1-q)\mathrm{Ln}_q(y_2)+1] \dots [(1-q)\mathrm{Ln}_q(y_h)+1] - 1}{1-q} + \right.\right. \\ \left.\left. 1\right][(1-q)\mathrm{Ln}_q(y_{h+1}) + 1] - 1\right\}, \quad \text{(B.42)}$$

after cross canceling $(1 - q)$ in the fraction in the first bracket,

$$\mathrm{Ln}_q(y_1) \oplus_q \dots \oplus_q \mathrm{Ln}_q(y_h) \oplus_q \mathrm{Ln}_q(y_{h+1}) = \\ = \frac{1}{1-q}\left\{\left[[(1-q)\mathrm{Ln}_q(y_1) + 1][(1-q)\mathrm{Ln}_q(y_2) + 1] \dots [(1-q)\mathrm{Ln}_q(y_h) + \right.\right. \\ \left.\left. 1] - 1 + 1\right][(1-q)\mathrm{Ln}_q(y_{h+1}) + 1] - 1\right\}, \quad \text{(B.43)}$$

and after canceling $-1 + 1$, it follows that

$$\mathrm{Ln}_q(y_1) \oplus_q \dots \oplus_q \mathrm{Ln}_q(y_h) \oplus_q \mathrm{Ln}_q(y_{h+1}) =$$

$$= \frac{[(1-q)\mathrm{Ln}_q(y_1)+1][(1-q)\mathrm{Ln}_q(y_2)+1]\ldots[(1-q)\mathrm{Ln}_q(y_h)+1][(1-q)\mathrm{Ln}_q(y_{h+1})+1]-1}{1-q} \quad \text{(B.44)}$$

quod erat demonstrandum.

On the other hand, for $q = 1$, we have

$$\mathrm{Ln}_1(y_1 y_2 \ldots y_n) = \mathrm{Ln}_1(y_1) \oplus_1 \mathrm{Ln}_1(y_2) \oplus_1 \ldots \oplus_1 \mathrm{Ln}_1(y_n)\,, \quad \text{(B.45)}$$

after considering $\oplus_1 \equiv +$, we obtain

$$\mathrm{Ln}(y_1 y_2 \ldots y_n) = \mathrm{Ln}(y_1) + \mathrm{Ln}(y_2) + \cdots + \mathrm{Ln}(y_n)\,, \quad \text{(B.46)}$$

which is the expected formula.

Finding Out a Formula for the *q*-Logarithm of the *q*-Product of Two Functions

We will search for a formula from the definition of this q-logarithm:

$$\mathrm{Ln}_q(x \otimes_q y) = \frac{(x \otimes_q y)^{1-q} - 1}{1-q}\,, \quad \text{(B.47)}$$

and by bearing in mind the definition of the q-product

$$a \otimes_q b = \begin{cases} [a^{1-q} + b^{1-q} - 1]^{\frac{1}{1-q}}, \text{if } a > 0, b > 0 \text{ and } [a^{1-q} + b^{1-q} - 1] > 0 \\ 0, \quad \text{elsewhere} \end{cases} \quad \text{(B.48)}$$

we have the following expression:

$$\mathrm{Ln}_q(x \otimes_q y) = \frac{\left[[a^{1-q}+b^{1-q}-1]^{\frac{1}{1-q}}\right]^{1-q} - 1}{1-q}\,, \quad \text{(B.49)}$$

by applying the power to power rule, it becomes thus:

$$\mathrm{Ln}_q(x \otimes_q y) = \frac{x^{1-q} + y^{1-q} - 1 - 1}{1-q}\,, \quad \text{(B.50)}$$

by separating terms, it is stated as

$$\mathrm{Ln}_q(x \otimes_q y) = \frac{x^{1-q} - 1}{1-q} + \frac{y^{1-q} - 1}{1-q}\,, \quad \text{(B.51)}$$

we use again the definition of the q-logarithm and obtain this way:

$$\mathrm{Ln}_q(x \otimes_q y) = \mathrm{Ln}_q(x) + \mathrm{Ln}_q(y)\,, \quad \text{(B.52)}$$

quod erat demonstrandum.

Furthermore, for $q = 1$, we have

$$\mathrm{Ln}_1(x \otimes_1 y) = \mathrm{Ln}_1(x) + \mathrm{Ln}_1(y)\,, \quad \text{(B.53)}$$

after considering $\otimes_1 \equiv \cdot$, we have as follows:

$$\mathrm{Ln}(x \cdot y) = \mathrm{Ln}(xy) = \mathrm{Ln}(x) + \mathrm{Ln}(y)\,, \quad \text{(B.54)}$$

which is the expected result.

Finding Out a Formula for the *q*-Logarithm of the Division of Two Functions

We start by using the definition of the q-logarithm. Then, we get this way:

$$\mathrm{Ln}_q\left(\frac{x}{y}\right) = \frac{1}{1-q}\left[\left(\frac{x}{y}\right)^{1-q} - 1\right], \qquad \text{(B.55)}$$

by applying the power of a quotient property and summing inside the brackets, we have

$$\mathrm{Ln}_q\left(\frac{x}{y}\right) = \frac{1}{1-q}\left(\frac{x^{1-q}}{y^{1-q}} - 1\right) = \frac{1}{1-q}\left(\frac{x^{1-q} - y^{1-q}}{y^{1-q}}\right); \quad \text{(B.56)}$$

next, we will use the following formulae:

$$x^{1-q} = (1-q)\mathrm{Ln}_q(x) + 1 \quad \text{and} \quad y^{1-q} = (1-q)\mathrm{Ln}_q(y) + 1 \text{ ,(B.57)}$$

then

$$\mathrm{Ln}_q\left(\frac{x}{y}\right) = \frac{1}{1-q}\left[\frac{(1-q)\mathrm{Ln}_q(x) + 1 - [(1-q)\mathrm{Ln}_q(y) + 1]}{(1-q)\mathrm{Ln}_q(y) + 1}\right], \qquad \text{(B.58)}$$

after operating in the numerator,

$$\mathrm{Ln}_q\left(\frac{x}{y}\right) = \frac{1}{1-q}\left[\frac{(1-q)\mathrm{Ln}_q(x) - (1-q)\mathrm{Ln}_q(y)}{(1-q)\mathrm{Ln}_q(y)+1}\right], \qquad \text{(B.59)}$$

and, after cross canceling $(1-q)$,

$$\mathrm{Ln}_q\left(\frac{x}{y}\right) = \frac{\mathrm{Ln}_q(x) - \mathrm{Ln}_q(y)}{1+(1-q)\mathrm{Ln}_q(y)} \ . \qquad \text{(B.60)}$$

We realize the last formula corresponds to the q-difference of $\mathrm{Ln}_q(x)$ and $\mathrm{Ln}_q(y)$, that is,

$$\mathrm{Ln}_q\left(\frac{x}{y}\right) = \mathrm{Ln}_q(x) \ominus_q \mathrm{Ln}_q(y) \ . \qquad \text{(B.61)}$$

which is the result we were looking for.

Furthermore, for $q = 1$, we have

$$\mathrm{Ln}_1\left(\frac{x}{y}\right) = \mathrm{Ln}_1(x) \ominus_1 \mathrm{Ln}_1(y) \ , \qquad \text{(B.62)}$$

after applying $\ominus_1 \equiv -$, we have the following result:

$$\mathrm{Ln}\left(\frac{x}{y}\right) = \mathrm{Ln}(x) - \mathrm{Ln}(y) \ , \qquad \text{(B.63)}$$

which is the expected result.

Finding Out a Formula for the *q*-Logarithm of the *q*-Division of Two Functions

We begin with the expression that defines the q-logarithm:

$$\mathrm{Ln}_q(x \oslash_q y) = \frac{1}{1-q}\left[(x \oslash_q y)^{1-q} - 1\right]; \qquad \text{(B.64)}$$

next, we use the definition of the q-division, $a \oslash_q b = [a^{1-q} - b^{1-q} + 1]^{\frac{1}{1-q}}$ and obtain as follows:

$$\mathrm{Ln}_q\left(x \oslash_q y\right) = \frac{1}{1-q}\left[\left(\left[x^{1-q} - y^{1-q} + 1\right]^{\frac{1}{1-q}}\right)^{1-q} - 1\right], \quad \text{(B.65)}$$

now, we will apply the power to power rule and obtain this way:

$$\mathrm{Ln}_q\left(x \oslash_q y\right) = \frac{1}{1-q}\left[x^{1-q} - y^{1-q} + 1 - 1\right], \quad \text{(B.66)}$$

after changing the order of the addends, it follows that

$$\mathrm{Ln}_q\left(x \oslash_q y\right) = \frac{1}{1-q}\left[x^{1-q} - 1 - y^{1-q} + 1\right] = \frac{x^{1-q}-1}{1-q} - \frac{y^{1-q}-1}{1-q}, \quad \text{(B.67)}$$

then, we use again the definition of the q-logarithm, and the following formula holds:

$$\mathrm{Ln}_q\left(x \oslash_q y\right) = \mathrm{Ln}_q(x) - \mathrm{Ln}_q(y)\,. \quad \text{(B.68)}$$

Furthermore, for $q = 1$, we obtain thus:

$$\mathrm{Ln}_1(x \oslash_1 y) = \mathrm{Ln}_1(x) - \mathrm{Ln}_1(y)\,, \quad \text{(B.69)}$$

by considering $\oslash_1 \equiv /$ and $\mathrm{Ln}_1(a) \equiv \mathrm{Ln}(a)$, we arrive at

$$\mathrm{Ln}(x/y) = \mathrm{Ln}(x) - \mathrm{Ln}(y)\,, \quad \text{(B.70)}$$

which is the expected result.

APPENDIX C

DEMONSTRATION OF SEVERAL PROPERTIES OF THE Q-OPERATIONS AND Q-GAUSSIAN

Finding Out Properties of the *q*-Sum

We must point out that there is no distributive property of the q-sum. Nonetheless, we will deduce other properties.

Finding Out the Neutral Element for the q-Sum

We will verify if 0 is the neutral element for the q-sum. Thus, according to the definition of the q-sum,

$$x \oplus_q 0 = x + 0 + (1-q)x \cdot 0 = x\,, \quad \text{(C.1)}$$

quod erat demonstrandum.

Furthermore, for $q = 1$, we have

$$x \oplus_1 0 = x + 0 + (1-1)x \cdot 0 = x\,, \quad \text{(C.2)}$$

$$x + 0 = x\,, \quad \text{(C.3)}$$

which is the expected result.

Finding Out the Additive Inverse for the q-Sum

We search for a number I_a satisfying the following formula:

$$x \oplus_q I_{\mathrm{a}} = 0. \quad \text{(C.4)}$$

Then, we will use the definition of the q-sum,

$$x \oplus_q I_{\mathrm{a}} = x + I_{\mathrm{a}} + (1-q)xI_{\mathrm{a}} = 0\,, \quad \text{(C.5)}$$

solving for I_{a}, it follows that

$$x + I_{\mathrm{a}}[1 + (1-q)x] = 0, \quad \text{(C.6)}$$

$$I_{\mathrm{a}} = \frac{-x}{1+(1-q)x} \quad \text{with} \quad [1 + (1-q)x] \neq 0\,; \quad \text{(C.7)}$$

to keep the similarity with the standard notation, we will abandon the symbol I_{a} and state it thus:

$$\ominus_q x = \frac{-x}{1+(1-q)x} \quad \text{with} \quad [1 + (1-q)x] \neq 0\,. \quad \text{(C.8)}$$

For $q = 1$, we have

$$\ominus_1 x = \frac{-x}{1+(1-1)x} = -x\,, \qquad \text{(C.9)}$$

which is the expected result.

Furthermore, it must be valid the formula $\ominus_q x \oplus_q x = 0$. Next, we will prove it. Thereupon,

$$\ominus_q x \oplus_q x = (\ominus_q x)\oplus_q x = (\ominus_q x) + x + (1-q)(\ominus_q x)x\,, \quad \text{(C.10)}$$

where we have used the definition of the q-sum. Now, we will consider the definition of $\ominus_q x$,

$$\ominus_q x \oplus_q x = \frac{-x}{1+(1-q)x} + x + (1-q)\left[\frac{-x}{1+(1-q)x}\right]x\,, \text{(C.11)}$$

and after summing the corresponding terms, we have

$$\ominus_q x \oplus_q x = \frac{-x + x[1+(1-q)x] - x^2(1-q)}{1+(1-q)x} = \frac{(1-q)x^2 - x^2(1-q)}{1+(1-q)x}\,, \quad \text{(C.12)}$$

therefore,

$$\ominus_q x \oplus_q x = 0\,. \qquad \text{(C.13)}$$

For $q = 1$, it is trivial,

$$\ominus_1 x \oplus_1 x = -x + x = 0\,. \ \text{(C.14)}$$

Finding Out the Associative Property of the q-Sum

We will verify the formula $x_1 \oplus_q (x_2 \oplus_q x_3) = (x_1 \oplus_q x_2) \oplus_q x_3$. Then, on the left side, we start by using the definition of the q-sum,

$$x_1 \oplus_q (x_2 \oplus_q x_3) = x_1 + (x_2 \oplus_q x_3) + (1-q)x_1(x_2 \oplus_q x_3)\,; \quad \text{(C.15)}$$

now, by defining the auxiliary parameter $D_3 = x_1 \oplus_q (x_2 \oplus_q x_3)$ and using again the q-sum, we have

$$D_3 = x_1 + [x_2 + x_3 + (1-q)x_2x_3] + (1-q)x_1[x_2 + x_3 + (1-q)x_2x_3]\,, \text{(C.16)}$$

after eliminating the first pair of brackets, we have

$$D_3 = x_1 + x_2 + x_3 + (1-q)x_2x_3 + (1-q)x_1x_2 + (1-q)x_1[x_3 + (1-q)x_2x_3]\,, \text{(C.17)}$$

by grouping terms, we obtain this way:

$$D_3 = x_1 + x_2 + (1-q)x_1x_2 + x_3 + (1-q)x_2x_3 + (1-q)x_1x_3 + (1-q)x_1(1-q)x_2x_3\,; \quad \text{(C.18)}$$

now, we introduce the symbol of the q-sum for the three first terms; we also factor $(1-q)x_3$ out in the other terms,

$$D_3 = (x_1 \oplus_q x_2) + x_3 + (1-q)x_3[x_2 + x_1 + (1-q)x_1x_2]\,, \ \text{(C.19)}$$

we recognize another q-sum inside brackets; consequently,

$$D_3 = (x_1 \oplus_q x_2) + x_3 + (1-q)x_3[x_1 \oplus_q x_2]\,, \quad \text{(C.20)}$$

then, we will apply the commutative property to the last addend. Thus,

$$D_3 = (x_1 \oplus_q x_2) + x_3 + (1-q)[x_1 \oplus_q x_2]x_3\,, \quad \text{(C.21)}$$

and finally, we recognize once more another q-sum:

$$D_3 = (x_1 \oplus_q x_2) \oplus_q x_3\,, \qquad \text{(C.22)}$$

that is, we have demonstrated $x_1 \oplus_q (x_2 \oplus_q x_3) = (x_1 \oplus_q x_2) \oplus_q x_3$.

Furthermore, for $q = 1$, we have

$$x_1 \oplus_1 (x_2 \oplus_1 x_3) = (x_1 \oplus_1 x_2) \oplus_1 x_3 \,, \quad \text{(C.23)}$$

after using $\oplus_1 \equiv +$, we have

$$x_1 + (x_2 + x_3) = (x_1 + x_2) + x_3 \,, \quad \text{(C.24)}$$

which is an established result.

Finding Out a Formula for $x_1 \oplus_q x_2 \oplus_q \dots \oplus_q x_n$ When $x_1 = x_2 = \cdots = x_n = x$

For $n = 2$, by using the definition of the q-sum,

$$x \oplus_q x = x + x + (1 - q)xx = 2x + (1 - q)x^2 \,. \quad \text{(C.25)}$$

For $n = 3$, it is stated as follows:

$$x \oplus_q x \oplus_q x = \left(x \oplus_q x\right) \oplus_q x = x + (x \oplus_q x) + (1 - q)x(x \oplus_q x) \,, \quad \text{(C.26)}$$

after applying the result of Eq. (C.25), we have

$$x \oplus_q x \oplus_q x = x + [2x + (1 - q)x^2] + (1 - q)x[2x + (1 - q)x^2] \,, \quad \text{(C.27)}$$

and by operating on the right side, we obtain this way:

$$x \oplus_q x \oplus_q x = x + 2x + (1 - q)x^2 + (1 - q)2x^2 + (1 - q)^2 x^3 \,, \quad \text{(C.28)}$$

now, after summing common terms, we get

$$x \oplus_q x \oplus_q x = 3x + 3(1 - q)x^2 + (1 - q)^2 x^3. \quad \text{(C.29)}$$

Besides, for $n = 2$ and $n = 3$, we realize the results remind the square and cube of a binomial, respectively; however, it is necessary to carry out some previous steps.

Thereupon, for $n = 2$ we have

$$x \oplus_q x = 2x + (1 - q)x^2 = \frac{2(1-q)x + (1 - q)^2 x^2}{1-q} = \frac{2(1-q)x + (1 - q)^2 x^2 + 1 - 1}{1-q} \,, \quad \text{(C.30)}$$

after identifying $2(1 - q)x + (1 - q)^2 x^2 + 1$ as the square of a binomial, it follows that

$$x \oplus_q x = \frac{[(1-q)x + 1]^2 - 1}{1-q} \,, \quad \text{(C.31)}$$

Likewise, for $n = 3$, we have

$$x \oplus_q x \oplus_q x = 3x + 3(1 - q)x^2 + (1 - q)^2 x^3 = \frac{3(1 - q)x + 3(1 - q)^2 x^2 + (1 - q)^3 x^3}{1-q} \,, \quad \text{(C.32)}$$

adding +1 and subtracting -1

$$x \oplus_q x \oplus_q x = \frac{3(1 - q)x + 3(1 - q)^2 x^2 + (1 - q)^3 x^3 + 1 - 1}{1-q} \,, \quad \text{(C.33)}$$

and after identifying $3(1 - q)x + 3(1 - q)^2 x^2 + (1 - q)^3 x^3 + 1$ as the cube of a binomial, namely $[(1 - q)x + 1]^3$, we can express Eq. (C.33) as follows:

$$x \oplus_q x \oplus_q x = \frac{[(1-q)x + 1]^3 - 1}{1-q} \,. \quad \text{(C.34)}$$

Based on the results for $n = 2$ and $n = 3$, it is proposed that

$$x \oplus_q x \oplus_q \dots \oplus_q x = \frac{[(1-q)x + 1]^n - 1}{1-q} \,. \quad \text{(C.35)}$$

To demonstrate it, we will use the method of mathematical induction.

Thereupon, for $n = 2$, it has already been proved. Hence, for $n = h$, we will assume the following formula to be true:

$$x \oplus_q x \dots \oplus_q x = \frac{[(1-q)x+1]^h - 1}{1-q}. \quad \text{(C.36)}$$

For $n = h + 1$, it is stated as follows:

$$x \oplus_q x \dots \oplus_q x \oplus_q x = \frac{[(1-q)x+1]^{h+1} - 1}{1-q}. \quad \text{(C.37)}$$

then, by using the associative property and setting the auxiliary parameter $D_4 = (x \oplus_q x \dots \oplus_q x) \oplus_q x$, we have

$$D_4 = (x \oplus_q x \dots \oplus_q x) + x + (1-q)(x \oplus_q x \dots \oplus_q x)x, \quad \text{(C.38)}$$

and after considering the inductive hypothesis (for $n = h$), we get

$$D_4 = \frac{[(1-q)x+1]^h - 1}{1-q} + x + (1-q)\frac{[(1-q)x+1]^h - 1}{1-q}x, \quad \text{(C.39)}$$

on the right side, by summing the fractions with the variable x, it follows that

$$D_4 = \frac{[(1-q)x+1]^h - 1 + (1-q)x + (1-q)x\{[(1-q)x+1]^h - 1\}}{1-q}, \quad \text{(C.40)}$$

now, we operate and eliminate the braces in the numerator, then,

$$D_4 = \frac{[(1-q)x+1]^h - 1 + (1-q)x + (1-q)x[(1-q)x+1]^h - (1-q)x}{1-q}, \quad \text{(C.41)}$$

by canceling $(1-q)x$ and $-(1-q)x$, we have

$$D_4 = \frac{[(1-q)x+1]^h - 1 + (1-q)x[(1-q)x+1]^h}{1-q}, \quad \text{(C.42)}$$

setting $a = (1-q)x + 1$,

$$D_4 = \frac{a^h - 1 + (1-q)xa^h}{1-q} = \frac{a^h[1 + (1-q)x] - 1}{1-q}, \quad \text{(C.43)}$$

in the last numerator, we can identify a again. Thus,

$$D_4 = \frac{a^h a - 1}{1-q} = \frac{a^{h+1} - 1}{1-q}, \quad \text{(C.44)}$$

and after replacing the value of a,

$$D_4 = \frac{[(1-q)x+1]^{h+1} - 1}{1-q}, \quad \text{(C.45)}$$

that is, we have demonstrated Eq. (C.37). Consequently, we can affirm that Eq. (C.35) holds.

On the other hand, for $q = 1$, we apply $\oplus_1 \equiv +$. Thus, from Eq. (C.35), we obtain the following limit:

$$D_5 = x + \cdots + x = \lim_{q \to 1} \frac{[(1-q)x+1]^n - 1}{1-q}, \quad \text{(C.46)}$$

where we have introduced the auxiliary parameter $D_5 = x + \cdots + x$. Now, we realize we have the indeterminate form $\frac{0}{0}$. Then, we will use a rule invented by the French mathematician Guillaume François Antoine de

L'Hôpital. Thus, we differentiate the numerator and denominator respect to q,

$$D_5 = \lim_{q \to 1} \frac{n[(1-q)x+1]^{n-1}(-x)}{-1}, \qquad (C.47)$$

after cross canceling the minus sign, and after pulling nx out, too, we have

$$D_5 = nx \lim_{q \to 1} [(1-q)x+1]^{n-1}, \qquad (C.48)$$

also, we know the limit of a power is the power of the limit. Consequently,

$$D_5 = nx \left\{ \lim_{q \to 1} [(1-q)x+1] \right\}^{n-1}, \qquad (C.49)$$

finally

$$D_5 = nx, \qquad (C.50)$$

which implies the standard result $x + \cdots + x = nx$.

Finding Out Some Properties of the *q*-Difference

First of all, for the q-difference, there is no associative, commutative, or distributive property. Nonetheless, we will search for additional formulae.

Finding Out a Formula for $x_1 \ominus_q (x_2 \ominus_q x_3)$

We will use the definition of the q-difference, $a \ominus_q b = \frac{a-b}{1+(1-q)b}$ with $[1+(1-q)b] > 0$. Then, we have the following expression:

$$x_1 \ominus_q (x_2 \ominus_q x_3) = \frac{x_1 - (x_2 \ominus_q x_3)}{1+(1-q)(x_2 \ominus_q x_3)} \qquad (C.51)$$

by establishing the auxiliary parameter $D_6 = x_1 \ominus_q (x_2 \ominus_q x_3)$ and using again the above definition, we have

$$D_6 = \ominus_q (x_2 \ominus_q x_3) = \frac{x_1 - \left[\frac{x_2 - x_3}{1+(1-q)x_3}\right]}{1+(1-q)\left[\frac{x_2 - x_3}{1+(1-q)x_3}\right]} = \frac{\frac{x_1[1+(1-q)x_3] - x_2 + x_3}{1+(1-q)x_3}}{\frac{1+(1-q)x_3 + (1-q)(x_2 - x_3)}{1+(1-q)x_3}}, \qquad (C.52)$$

by cross canceling the corresponding denominators, it follows that

$$D_6 = \frac{x_1[1+(1-q)x_3] - x_2 + x_3}{1+(1-q)x_3 + (1-q)(x_2 - x_3)}, \qquad (C.53)$$

after operating,

$$D_6 = \frac{x_1 + (1-q)x_1 x_3 - x_2 + x_3}{1+(1-q)x_3 + (1-q)x_2 - (1-q)x_3}, \qquad (C.54)$$

now, we will cancel the opposite terms in the denominator. Thus, it becomes this way:

$$D_6 = \frac{x_1 + (1-q)x_1 x_3 - x_2 + x_3}{1+(1-q)x_2}, \qquad (C.55)$$

and by changing the order of the addends,

$$D_6 = \frac{x_1 - x_2 + (1-q)x_1 x_3 + x_3}{1+(1-q)x_2}, \qquad (C.56)$$

now, we will separate it in two fractions. Then,

$$D_6 = \frac{x_1 - x_2}{1 + (1-q)x_2} + \frac{(1-q)x_1x_3 + x_3}{1 + (1-q)x_2}, \qquad \text{(C.57)}$$

by adding and subtracting x_3, it follows that

$$D_6 = \frac{x_1 - x_2}{1 + (1-q)x_2} + x_3 - x_3 + \frac{(1-q)x_1x_3 + x_3}{1 + (1-q)x_2}, \quad \text{(C.58)}$$

and by multiplying by $\frac{1 + (1-q)x_2}{1 + (1-q)x_2}$ the third addend, we have the following expression:

$$D_6 = \frac{x_1 - x_2}{1 + (1-q)x_2} + x_3 - x_3\frac{1 + (1-q)x_2}{1 + (1-q)x_2} + \frac{(1-q)x_1x_3 + x_3}{1 + (1-q)x_2}, \text{(C.59)}$$

now, we sum the third and the fourth addends and obtain as follows:

$$D_6 = \frac{x_1 - x_2}{1 + (1-q)x_2} + x_3 + \frac{-x_3 - (1-q)x_2x_3 + (1-q)x_1x_3 + x_3}{1 + (1-q)x_2}, \quad \text{(C.60)}$$

by canceling $-x_3$ and x_3 in the third addend, it is stated as

$$D_6 = \frac{x_1 - x_2}{1 + (1-q)x_2} + x_3 + \frac{-(1-q)x_2x_3 + (1-q)x_1x_3}{1 + (1-q)x_2}, \qquad \text{(C.61)}$$

by factoring out $(1 - q)$ and $-x_3$ in the last addend, we obtain this way:

$$D_6 = \frac{x_1 - x_2}{1 + (1-q)x_2} + x_3 + (1 - q)\frac{x_1 - x_2}{1 + (1-q)x_2}x_3, \qquad \text{(C.62)}$$

and by using the definition of the q-difference, we get

$$D_6 = (x_1 \ominus_q x_2) + x_3 + (1 - q)(x_1 \ominus_q x_2)x_3, \qquad \text{(C.63)}$$

finally, on the right side, we realize we have a q-sum, that is,

$$D_6 = (x_1 \ominus_q x_2) \oplus_q x_3; \qquad \text{(C.64)}$$

consequently, the following expression holds: $x_1 \ominus_q (x_2 \ominus_q x_3) = (x_1 \ominus_q x_2) \oplus_q x_3$.

On the other hand, for $q = 1$, we recover the well-known formulae

$$x_1 - (x_2 - x_3) = (x_1 - x_2) + x_3 \text{ or } x_1 - (x_2 - x_3) = x_1 - x_2 + x_3. \text{(C.65)}$$

Finding Out an Additional Formula for $x \ominus_q y$

We start by using the definition of the q-difference. Thereupon,

$$x \ominus_q y = \frac{x - y}{1 + (1-q)y}, \qquad \text{(C.66)}$$

and after separating it in two fractions, it follows that

$$x \ominus_q y = \frac{-y}{1 + (1-q)y} + \frac{x}{1 + (1-q)y}, \qquad \text{(C.67)}$$

by adding and subtracting x, we have the following expression:

$$x \ominus_q y = \frac{-y}{1 + (1-q)y} + x - x + \frac{x}{1 + (1-q)y}, \quad \text{(C.68)}$$

operating on $-x$,

$$x \ominus_q = \frac{-y}{1 + (1-q)y} + x - \frac{x[1 + (1-q)y]}{1 + (1-q)y} + \frac{x}{1 + (1-q)y}, \quad \text{(C.69)}$$

after summing the last two terms, we obtain as follows:

$$x \ominus_q y = \frac{-y}{1+(1-q)y} + x + (1-q)\frac{-yx}{1+(1-q)y}\ ; \quad \text{(C.70)}$$

we can identify the q-additive inverse of y, that is, $\ominus_q y = \frac{-y}{1+(1-q)y}$.

Therefore,

$$x \ominus_q y = (\ominus_q y) + x + (1-q)(\ominus_q y)x\ ; \quad \text{(C.71)}$$

however, on the right side of the last equation, we recognize the q-sum of $(\ominus_q y)$ and x, namely,

$$x \ominus_q y = (\ominus_q y) \oplus_q x\ . \qquad \text{(C.72)}$$

Furthermore, we can get rid of the parentheses, and we obtain this way:

$$x \ominus_q y = \ominus_q y \oplus_q x\ . \qquad \text{(C.73)}$$

For $q = 1$, we have $x \ominus_1 y = \ominus_1 y \oplus_1 x$; and after taking into account $\ominus_1 \equiv -$ and $\oplus_1 \equiv +$, it is stated as $x - y = -y + x$, which is a standard result.

Finding Out Properties of the *q*-Product

Finding Out the Commutative Property of the q-Product

We will search for verifying the formula $x_1 \otimes_q x_2 = x_2 \otimes_q x_1$. Then, we will use the definition of the q-product on the left side of the previous formula. Thus, it follows that

$$x_1 \otimes_q x_2 = [x_1{}^{1-q} + x_2{}^{1-q} - 1]^{\frac{1}{1-q}} = [x_2{}^{1-q} + x_1{}^{1-q} - 1]^{\frac{1}{1-q}}, \quad \text{(C.74)}$$

now, we will again use the definition of the q-product, on the right side of the last formula, and obtain as follows:

$$x_1 \otimes_q x_2 = x_2 \otimes_q x_1\ , \qquad \text{(C.75)}$$

quod erat demonstrandum.

Finally, for $q = 1$, we have $x_1 \otimes_1 x_2 = x_2 \otimes_1 x_1$. Thus, after bearing in mind that $\otimes_1$ is equivalent to the usual multiplication, we recover the well-known formula

$$x_1 x_2 = x_2 x_1\ . \qquad \text{(C.76)}$$

Finding Out the Neutral Element for the q-Product

We will verify if 1 is the neutral element for the q-product. Then, according to the definition of the q-product,

$$x \otimes_q 1 = [x^{1-q} + 1^{1-q} - 1]^{\frac{1}{1-q}} = [x^{1-q}]^{\frac{1}{1-q}} = x\ , \qquad \text{(C.77)}$$

quod erat demonstrandum.

Finally, for $q = 1$, we recover the well-known formula

$$x \cdot 1 = x\ . \qquad \text{(C.78)}$$

Finding Out the Multiplicative Inverse for the q-Product

We search for a number I_m satisfying the following expression:

$$x \otimes_q I_m = 1\,, \qquad \text{(C.79)}$$

then, in accordance with the definition of the q-product, it follows that

$$x \otimes_q I_m = [x^{1-q} + (I_m)^{1-q} - 1]^{\frac{1}{1-q}}\,, \qquad \text{(C.80)}$$

and raising to the power $1 - q$ on the right side, we obtain as follows:

$$x^{1-q} + (I_m)^{1-q} - 1 = 1^{1-q} = 1\,, \qquad \text{(C.81)}$$

and solving for I_m, it becomes thus:

$$(I_m)^{1-q} = 1 + 1 - x^{1-q}\,, \qquad \text{(C.82)}$$

by taking the $(1 - q)$-th root, we have I_m is given by

$$I_m = (2 - x^{1-q})^{\frac{1}{1-q}}\,; \qquad \text{(C.83)}$$

now, in order to maintain similarity with the usual formulae, we must change the notation this way: $I_m \equiv 1 \oslash_q x$. Thus,

$$1 \oslash_q x = (2 - x^{1-q})^{\frac{1}{1-q}}\,. \qquad \text{(C.84)}$$

For $q = 1$, we have the indeterminate form 1^{∞}. Therefore, before dealing with $q = 1$, we will consider that the exponential and logarithm are inverse functions. Thus,

$$1 \oslash_q x = (2 - x^{1-q})^{\frac{1}{1-q}} = \exp\left\{\text{Ln}\left[(2 - x^{1-q})^{\frac{1}{1-q}}\right]\right\}, \qquad \text{(C.85)}$$

now, we use the logarithmic power rule and obtain thus:

$$1 \oslash_q x = \exp\left\{\frac{1}{1-q}\text{Ln}[2 - x^{1-q}]\right\}; \qquad \text{(C.86)}$$

Therefore, for $q = 1$, we have this expression:

$$1 \oslash_1 x = \mathbf{Lim}_{q\to 1}\left\{\exp\left[\frac{1}{1-q}\text{Ln}(2 - x^{1-q})\right]\right\}, \qquad \text{(C.87)}$$

by applying the properties of limit, we can interchange the limit and the exponential. Consequently, it follows that

$$1 \oslash_1 x = \exp\left\{\mathbf{Lim}_{q\to 1}\left[\frac{1}{1-q}\text{Ln}(2 - x^{1-q})\right]\right\}, \qquad \text{(C.88)}$$

the last equation is evidently an indeterminate form $\frac{0}{0}$. Then, we use L'Hôpital's rule, differentiating the numerator and denominator respect to q,

$$1 \oslash_1 x = \exp\left\{\mathbf{Lim}_{q\to 1}\left[\frac{\frac{-x^{1-q}(-\text{Ln}x)}{2-x^{1-q}}}{-1}\right]\right\} = \exp\left\{\mathbf{Lim}_{q\to 1}\left[\frac{-x^{1-q}\text{Ln}x}{2-x^{1-q}}\right]\right\}, \qquad \text{(C.89)}$$

after applying limit, we have this way:

$$1 \oslash_1 x = \exp\left\{\frac{-x^{1-1}\text{Ln}x}{2-x^{1-1}}\right\}, \qquad \text{(C.90)}$$

simplifying,

$$1 \oslash_1 x = \exp\left\{\mathrm{Ln}\left(\frac{1}{x}\right)\right\}, \qquad (C.91)$$

finally,

$$1 \oslash_1 x = \frac{1}{x}, \qquad (C.92)$$

which is the expected result. Evidently, the symbol $\oslash_1$ is the equivalent of the usual division sign, that is, $\oslash_1 \equiv /$.

Finding Out the Associative Property of the q-Product

We want to verify the formula $(x_1 \otimes_q x_2) \otimes_q x_3 = x_1 \otimes_q (x_2 \otimes_q x_3)$. Then, we will use the definition of the q-product on the left side of this expression. Besides, we will also set the following auxiliary parameter: $D_7 = (x_1 \otimes_q x_2) \otimes_q x_3$. Thus, it follows that

$$D_7 = \left[(x_1 \otimes_q x_2)^{1-q} + {x_3}^{1-q} - 1\right]^{\frac{1}{1-q}}; \qquad (C.93)$$

by using again the definition of the q-product inside the parentheses on the right side of last formula, it follows that

$$D_7 = \left[\left\{[{x_1}^{1-q} + {x_2}^{1-q} - 1]^{\frac{1}{1-q}}\right\}^{1-q} + {x_3}^{1-q} - 1\right]^{\frac{1}{1-q}}, \qquad (C.94)$$

now, inside the inner brackets, we will apply the power of a power property. Thus,

$$D_7 = [{x_1}^{1-q} + {x_2}^{1-q} - 1 + {x_3}^{1-q} - 1]^{\frac{1}{1-q}}, \qquad (C.95)$$

rearranging terms, we have

$$D_7 = [{x_1}^{1-q} + {x_2}^{1-q} + {x_3}^{1-q} - 1 - 1]^{\frac{1}{1-q}}, \qquad (C.96)$$

and separating terms conveniently, it follows that

$$D_7 = \left[{x_1}^{1-q} + \left\{[{x_2}^{1-q} + {x_3}^{1-q} - 1]^{\frac{1}{1-q}}\right\}^{1-q} - 1\right]^{\frac{1}{1-q}}, \qquad (C.97)$$

now, in the braces we will use the definition of the q-product.

Then, we obtain

$$D_7 = \left[{x_1}^{1-q} + (x_2 \otimes_q x_3)^{1-q} - 1\right]^{\frac{1}{1-q}}, \qquad (C.98)$$

and after using again the definition of the q-product on the right side, we arrive at

$$D_7 = x_1 \otimes_q (x_2 \otimes_q x_3). \qquad (C.99)$$

Therefore, we conclude that the q-product fulfils the associative property. Thereupon, we can establish the following set of formulae:

$$x_1 \otimes_q x_2 \otimes_q x_3 = (x_1 \otimes_q x_2) \otimes_q x_3 = x_1 \otimes_q (x_2 \otimes_q x_3). \qquad (C.100)$$

Finally, for $q = 1$, we have $(x_1 \otimes_1 x_2) \otimes_1 x_3 = x_1 \otimes_1 (x_2 \otimes_1 x_3)$. Consequently, after remembering that the symbol $\otimes_1$ is the equivalent of the usual multiplication sign, we recover the well-known formulae

$$x_1x_2x_3 = (x_1x_2)x_3 = x_1(x_2x_3)\,. \qquad \text{(C.101)}$$

Finding Out a Formula for $x_1 \otimes_q x_2 \otimes_q \dots \otimes_q x_n$ When $x_1 = x_2 = \cdots = x_n = x$

For $n = 2$, we start by using the definition of the q-product. Hence,

$$x \otimes_q x = [x^{1-q} + x^{1-q} - 1]^{\frac{1}{1-q}} = [2x^{1-q} - 1]^{\frac{1}{1-q}}\,. \qquad \text{(C.102)}$$

For $n = 3$, we will use the associative property of the q-product; consequently,

$$x \otimes_q x \otimes_q x = (x \otimes_q x) \otimes_q x = \left[x^{1-q} + (x \otimes_q x)^{1-q} - 1\right]^{\frac{1}{1-q}}, \qquad \text{(C.103)}$$

where we have used the definition of the q-product again. After using it once more, it becomes thus:

$$x \otimes_q x \otimes_q x = \left[x^{1-q} + \left([x^{1-q} + x^{1-q} - 1]^{\frac{1}{1-q}}\right)^{1-q} - 1\right]^{\frac{1}{1-q}}, \qquad \text{(C.104)}$$

Now, we simplify the second term in the outer brackets and obtain as follows:

$$x \otimes_q x \otimes_q x = [x^{1-q} + x^{1-q} + x^{1-q} - 1 - 1]^{\frac{1}{1-q}} = [3x^{1-q} - 2]^{\frac{1}{1-q}}; \qquad \text{(C.105)}$$

then, for $n = 2$ and $n = 3$, the previous results can be expressed, respectively, this way:

$$x \otimes_q x = [2x^{1-q} - (2-1)]^{\frac{1}{1-q}} \text{ and } x \otimes_q x \otimes_q x = [3x^{1-q} - 2]^{\frac{1}{1-q}}\,. \qquad \text{(C.106)}$$

Consequently, we can generalize and propose the following formula for any natural number n:

$$x \otimes_q x \otimes_q \dots \otimes_q x = [nx^{1-q} - (n-1)]^{\frac{1}{1-q}}, \qquad \text{(C.107)}$$

now, to demonstrate it, we will use the method of mathematical induction.

For $n = 2$, it has already been proven. For $n = h$, we formulate thus:

$$x \otimes_q x \otimes_q \dots \otimes_q x = [hx^{1-q} - (h-1)]^{\frac{1}{1-q}}, \qquad \text{(C.108)}$$

according to the aforementioned method, we will assume that the last equation is true.

For $n = h + 1$, it becomes this way:

$$x \otimes_q x \otimes_q \dots \otimes_q x \otimes_q x = \left[(h+1)x^{1-q} - ((h+1)-1)\right]^{\frac{1}{1-q}}; \qquad \text{(C.109)}$$

next, we use the associative property and set the following auxiliary parameter: $D_8 = (x \otimes_q x \otimes_q \dots \otimes_q x) \otimes_q x$. Hence,

$$D_8 = \left[(h+1)x^{1-q} - ((h+1)-1)\right]^{\frac{1}{1-q}}, \qquad \text{(C.110)}$$

so, to prove the expression on the right side, we use the q-product of $(x \otimes_q x \otimes_q \dots \otimes_q x)$ and x,

$$D_8 = \left[\left(x \otimes_q x \otimes_q \dots \otimes_q x\right)^{1-q} + x^{1-q} - 1\right]^{\frac{1}{1-q}}, \quad \text{(C.111)}$$

then, in the inner parentheses in Eq. (C.111), we use the inductive hypothesis. Thus,

$$D_8 = \left[\left([hx^{1-q} - (h-1)]^{\frac{1}{1-q}}\right)^{1-q} + x^{1-q} - 1\right]^{\frac{1}{1-q}}, \quad \text{(C.112)}$$

now, after operating and eliminating the outer parentheses and the inner brackets, it follows that $D_8 = [hx^{1-q} - (h-1) + x^{1-q} - 1]^{\frac{1}{1-q}}$, and by factoring out x^{1-q}, we have as follows:

$$D_8 = [(h+1)x^{1-q} - (h-1) - 1]^{\frac{1}{1-q}}, \quad \text{(C.113)}$$

now, we rearrange $-(h-1) - 1$ as $-\big((h+1) - 1\big)$. Then,

$$D_8 = \big[(h+1)x^{1-q} - \big((h+1) - 1\big)\big]^{\frac{1}{1-q}}, \quad \text{(C.114)}$$

thus, we have proved Eq. (C.110) and, consequently, also, $x \otimes_q x \otimes_q \dots \otimes_q x = [nx^{1-q} - (n-1)]^{\frac{1}{1-q}}$.

Furthermore, on the other hand, for $q = 1$, we have the following limit:

$$x \dots xx = \lim_{q \to 1} [nx^{1-q} - (n-1)]^{\frac{1}{1-q}}, \quad \text{(C.115)}$$

by factoring out n, we obtain

$$x \dots xx = \lim_{q \to 1} [n(x^{1-q} - 1) + 1]^{\frac{1}{1-q}}, \quad \text{(C.116)}$$

and after multiplying n by $\frac{1-q}{1-q}$, we obtain this way:

$$x \dots xx = \lim_{q \to 1} \left[n(1-q)\frac{x^{1-q} - 1}{1-q} + 1\right]^{\frac{1}{1-q}} = \lim_{q \to 1} \left[n(1-q)\mathrm{Ln}_q x + 1\right]^{\frac{1}{1-q}}, \quad \text{(C.117)}$$

where we utilized the definition of the q-logarithm. After changing the order of the addends, we get

$$x \dots xx = \lim_{q \to 1} \left[1 + n(1-q)\mathrm{Ln}_q x\right]^{\frac{1}{1-q}}, \quad \text{(C.118)}$$

now, we realize it is the limit of a q-exponential function, specifically,

$$x \dots xx = \lim_{q \to 1} \exp_q(n\mathrm{Ln}_q x), \quad \text{(C.119)}$$

by applying limit, we obtain the well-known formulae

$$x \dots xx = \exp(n\mathrm{Ln}x) = \exp(\mathrm{Ln}(x^n)) = x^n, \quad \text{(C.120)}$$

which is a standard result.

Finding Out Properties of the q-Division

Finding Out a Formula for $x_1 \oslash_q (x_2 \oslash_q x_3)$

We start by applying the definition of the q-division, that is, $a \oslash_q b = [a^{1-q} - b^{1-q} + 1]^{\frac{1}{1-q}}$ with $[a^{1-q} - b^{1-q} + 1] \geq 0$. Thus,

$$x_1 \oslash_q (x_2 \oslash_q x_3) = x_1 \oslash_q \left([{x_2}^{1-q} - {x_3}^{1-q} + 1]^{\frac{1}{1-q}}\right), \quad \text{(C.121)}$$

now, we set the auxiliary parameter $D_9 = x_1 \oslash_q (x_2 \oslash_q x_3)$ and after applying again the aforementioned definition, we obtain

$$D_9 = \left[{x_1}^{1-q} - \left\{[{x_2}^{1-q} - {x_3}^{1-q} + 1]^{\frac{1}{1-q}}\right\}^{1-q} + 1\right]^{\frac{1}{1-q}}, \quad \text{(C.122)}$$

and by canceling the exponents around the right brace, we have as follows:

$$D_9 = [{x_1}^{1-q} - [{x_2}^{1-q} - {x_3}^{1-q} + 1] + 1]^{\frac{1}{1-q}}, \quad \text{(C.123)}$$

after operating and deleting the inner brackets, it follows that

$$D_9 = [{x_1}^{1-q} - {x_2}^{1-q} + {x_3}^{1-q} - 1 + 1]^{\frac{1}{1-q}}, \quad \text{(C.124)}$$

rearranging terms,

$$D_9 = \left[\left\{[{x_1}^{1-q} - {x_2}^{1-q} + 1]^{\frac{1}{1-q}}\right\}^{1-q} + {x_3}^{1-q} - 1\right]^{\frac{1}{1-q}}, \quad \text{(C.125)}$$

now, we realize it is not possible to obtain the fictitious formula $x_1 \oslash_q (x_2 \oslash_q x_3) = (x_1 \oslash_q x_2) \oslash_q x_3$. However, in the last numbered equation, we perceive we have the q-division between x_1 and x_2, that is,

$$D_9 = [\,(x_1 \oslash_q x_2)^{1-q} + {x_3}^{1-q} - 1]^{\frac{1}{1-q}}, \quad \text{(C.126)}$$

now, we clearly recognize the q-product between $x_1 \oslash_q x_2$ and x_3; therefore,

$$D_9 = (x_1 \oslash_q x_2) \otimes_q x_3 \,. \quad \text{(C.127)}$$

Furthermore, in Eq. (C.124) we also can group the terms in another way, namely,

$$D_9 = [{x_1}^{1-q} + {x_3}^{1-q} - 1 - {x_2}^{1-q} + 1]^{\frac{1}{1-q}}, \quad \text{(C.128)}$$

and manipulating the three first addends on the last brackets, it follows that

$$D_9 = \left[\left\{[{x_1}^{1-q} + {x_3}^{1-q} - 1]^{\frac{1}{1-q}}\right\}^{1-q} - {x_2}^{1-q} + 1\right]^{\frac{1}{1-q}}, \quad \text{(C.129)}$$

now, by using the definition of the q-product, we have as follows:

$$D_9 = \left[\left(x_1 \otimes_q x_3\right)^{1-q} - {x_2}^{1-q} + 1\right]^{\frac{1}{1-q}}, \quad \text{(C.130)}$$

so, on the right side we can identify a q-division, that is, $D_9 = (x_1 \otimes_q x_3) \oslash_q x_2$. Thus, we can establish

$$x_1 \oslash_q (x_2 \oslash_q x_3) = (x_1 \otimes_q x_3) \oslash_q x_2 \ . \qquad \text{(C.131)}$$

On the other hand, for $q = 1$, after considering that $\oslash_q$ is the equivalent of the usual division sign, we recover the following well-known formulae:

$$x_1/(x_2/x_3) = (x_1 x_3)/x_2 \qquad \text{or} \qquad \frac{x_1}{\frac{x_2}{x_3}} = \frac{x_1 x_3}{x_2}, \qquad \text{(C.132)}$$

which are obviously standard formulae.

Finding Out a Formula for $1 \oslash_q (1 \oslash_q x)$

We will use the definition of the multiplicative inverse of the q-product, $1 \oslash_q a = (2 - a^{1-q})^{\frac{1}{1-q}}$. Then,

$$1 \oslash_q (1 \oslash_q x) = \left[2 - (1 \oslash_q x)^{1-q}\right]^{\frac{1}{1-q}}, \qquad \text{(C.133)}$$

on the right side, we use again the definition of the multiplicative inverse of the q-product,

$$1 \oslash_q (1 \oslash_q x) = \left\{2 - \left[(2 - x^{1-q})^{\frac{1}{1-q}}\right]^{1-q}\right\}^{\frac{1}{1-q}}, \qquad \text{(C.134)}$$

after canceling exponents around the right bracket, we have thus:

$$1 \oslash_q (1 \oslash_q x) = \{2 - (2 - x^{1-q})\}^{\frac{1}{1-q}} = \{x^{1-q}\}^{\frac{1}{1-q}}, \qquad \text{(C.135)}$$

simplifying

$$1 \oslash_q (1 \oslash_q x) = x\ , \qquad \text{(C.136)}$$

which is the expected result.

On the other hand, for $q = 1$, it results that

$$1 \oslash_1 (1 \oslash_1 x) = 1/(1/x) = x\ . \qquad \text{(C.137)}$$

Finding Out a Formula for $1 \oslash_q (x \oslash_q y)$

From the formula $x_1 \oslash_q (x_2 \oslash_q x_3) = (x_1 \otimes_q x_3) \oslash_q x_2$, let $x_1 \equiv 1$, $x_2 \equiv x$, and $x_3 \equiv y$. Then, we obtain the following expression:

$$1 \oslash_q (x \oslash_q y) = (1 \otimes_q y) \oslash_q x\ , \qquad \text{(C.138)}$$

now, we take into account that $1 \otimes_q y = y$ in accordance with the neutral element of the q-product. Therefore, it follows that

$$1 \oslash_q (x \oslash_q y) = y \oslash_q x\ . \qquad \text{(C.139)}$$

On the other hand, for $q = 1$, we have $1 \oslash_1 (x \oslash_1 y) = y \oslash_1 x$, which can be expressed as $1/(x/y) = y/x$. This is obviously a known formula.

Finding Out a Formula for $1 \oslash_q 0$

From the definition of the multiplicative inverse of the q-product, we formulate this way:

$$1 \oslash_q 0 = (2 - 0^{1-q})^{\frac{1}{1-q}}\,, \qquad \text{(C.140)}$$

after operating, we obtain this way:

$$1 \oslash_q 0 = \begin{cases} 2^{\frac{1}{1-q}}, & q < 1 \\ +\infty, & q \geq 1 \end{cases}\,, \qquad \text{(C.141)}$$

that is, $1 \oslash_q 0$ does not diverge for $q < 1$, which is an interesting result.

Demonstrations Concerning the Standard Gaussian Distribution

Demonstrating How to Obtain the Standard Gaussian Distribution From Differential Entropy

We will use differential entropy, which is used in the theory of information. This entropy is not a simple extension of the similar discrete entropy. As opposed to the discrete case, where the entropy is associated with the quantum states of a given system, differential or continuous entropy has no immediate physical interpretation.

By setting the probability distribution $p(x)] \equiv N(x)$, differential entropy is defined by

$$h[N(x)] = -\int_x N(x)\mathrm{Ln}(N(x))\,dx\,, \qquad \text{(C.142)}$$

now, we will use the maximum entropy principle along with the Lagrange multipliers. Then, we will consider two constraints: normalization condition and requirement of a fixed variance, respectively,

$$\int_{-\infty}^{+\infty} N(x)\,dx = 1 \quad \text{and} \quad \sigma^2 = \int_{-\infty}^{+\infty} N(x)(x-\mu)^2\,dx\,; \qquad \text{(C.143)}$$

where σ^2 stands for the standard deviation and μ is the mean defined as $\mu = \int_{-\infty}^{+\infty} N(x)x\,dx$. This mean is equivalent to the first order moment $E[x]$ of the Gaussian distribution. Likewise, we set the Lagrangian function

$$L = h[N(x)] - \alpha \int_{-\infty}^{+\infty} N(x)\,dx - \beta \int_{-\infty}^{+\infty} N(x)(x-\mu)^2\,dx\,; \qquad \text{(C.144)}$$

next, we will differentiate the Lagrangian function respect to $N(x)$, and we will equalize it to zero:

$$\frac{\partial L}{\partial N}=\frac{\partial}{\partial N}\left\{h[N(x)]-\alpha\int_{-\infty}^{+\infty}N(x)\,dx-\beta\int_{-\infty}^{+\infty}N(x)(x-\mu)^2\,dx\right\}=0\,,\quad\text{(C.145)}$$

after replacing differential entropy, we have this way:

$$\frac{\partial L}{\partial N}=\frac{\partial}{\partial N}\{-\int_{-\infty}^{+\infty}N(x)\mathrm{Ln}(N(x))\,dx-\alpha\int_{-\infty}^{+\infty}N(x)\,dx-\beta\int_{-\infty}^{+\infty}N(x)(x-\mu)^2\,dx\}=0,\text{(C.146)}$$

and after operating on the corresponding integrands,

$$\frac{\partial L}{\partial N}=\int_{-\infty}^{+\infty}\{-\mathrm{Ln}(N(x))-1-\alpha-\beta(x-\mu)^2\}\,dx=0\,,\quad\text{(C.147)}$$

now, the integrand must be zero,

$$-\mathrm{Ln}(N(x))-1-\alpha-\beta(x-\mu)^2=0\,,\quad\text{(C.148)}$$

after solving for $\mathrm{Ln}(N(x))$, it follows that

$$\mathrm{Ln}(N(x))=-1-\alpha-\beta(x-\mu)^2\,,\quad\text{(C.149)}$$

taking the exponential on both sides, and considering that the exponential function and the logarithm function are inverse functions, we have

$$N(x)=\exp(-1-\alpha-\beta(x-\mu)^2)\,;\quad\text{(C.150)}$$

putting outside the constant factor $\exp(-1-\alpha)$, we obtain this way:

$$N(x)=\exp(-1-\alpha)\exp(-\beta(x-\mu)^2)\,,\quad\text{(C.151)}$$

integrating both sides from $-\infty$ to $+\infty$,

$$\int_{-\infty}^{+\infty}N(x)\,dx=\exp(-1-\alpha)\int_{-\infty}^{+\infty}\exp(-\beta(x-\mu)^2)\,dx,\quad\text{(C.152)}$$

now, on the left side, we will use the normalization condition; and, on the right side, we will use $\int_{-\infty}^{+\infty}\exp(-\beta(x-\mu)^2)\,dx=\sqrt{\frac{\pi}{\beta}}$. Thus,

$$1=\exp(-1-\alpha)\sqrt{\frac{\pi}{\beta}},\quad\text{(C.153)}$$

then, we have as follows:

$$\exp(-1-\alpha)=\sqrt{\frac{\beta}{\pi}},\quad\text{(C.154)}$$

therefore, by replacing Eq. (C.154) into Eq. (C.151), we arrive at the formula

$$N(x)=\sqrt{\frac{\beta}{\pi}}\exp(-\beta(x-\mu)^2)\,.\quad\text{(C.155)}$$

Besides, to determine the value of β, we multiply Eq. (C.155) by $(x-\mu)^2$ and we integrate both sides from $-\infty$ to $+\infty$. Then, it follows that

$$\int_{-\infty}^{+\infty}(x-\mu)^2N(x)\,dx=\sqrt{\frac{\beta}{\pi}}\int_{-\infty}^{+\infty}(x-\mu)^2\exp(-\beta(x-\mu)^2)\,dx\,;\quad\text{(C.156)}$$

thereupon, on the left side, we can identify the standard deviation $\int_{-\infty}^{+\infty}(x-\mu)^2N(x)\,dx=\sigma^2$. Therefore,

$$\sigma^2 = \sqrt{\frac{\beta}{\pi}} \int_{-\infty}^{+\infty} (x-\mu)^2 \exp(-\beta(x-\mu)^2)\, dx \; ; \qquad \text{(C.157)}$$

or

$$\sigma^2\sqrt{\pi} = \sqrt{\beta} \int_{-\infty}^{+\infty} (x-\mu)^2 \exp(-\beta(x-\mu)^2)\, dx \; ; \qquad \text{(C.158)}$$

now, we will set $z = \sqrt{\beta}\,(x-\mu)$ and $dz = \sqrt{\beta}\, dx$. Thus,

$$\sigma^2\sqrt{\pi} = \frac{1}{\beta} \int_{-\infty}^{+\infty} z^2 \exp(-z^2)\, dz \; ; \qquad \text{(C.159)}$$

after evaluating the integral, it results in $\int_{-\infty}^{+\infty} z^2 \exp(-z^2)\, dz = \frac{\sqrt{\pi}}{2}$. Consequently,

$$\sigma^2\sqrt{\pi} = \frac{1}{\beta}\frac{\sqrt{\pi}}{2} \; ; \qquad \text{(C.160)}$$

we will solve for β and, afterward, we will find its square root, that is,

$$\beta = \frac{1}{2\sigma^2} \quad \text{and} \quad \sqrt{\beta} = \frac{1}{\sqrt{2}\sigma} \; ; \qquad \text{(C.161)}$$

then, replacing $\sqrt{\beta} = \frac{1}{\sqrt{2}\sigma}$ into Eq. (C.155), the probability distribution is given by

$$N(x) = \sqrt{\frac{1}{2\sigma^2\pi}} \exp\left(-\frac{1}{2\sigma^2}\,(x-\mu)^2\right) \; ; \qquad \text{(C.162)}$$

or

$$N(x) = \frac{1}{\sqrt{2\pi}\sigma} \exp\left[-\frac{1}{2}\left(\frac{x-\mu}{\sigma}\right)^2\right] ; \qquad \text{(C.163)}$$

quod erat demonstrandum. This final formula is the well-known expression for the standard Gaussian distribution.

Demonstrating How to Obtain the Standard Kurtosis for the Standard Gaussian distribution

Kurtosis was a breakthrough carried out by the English mathematician and statistician Karl Pearson in 1905. It is a parameter that is used to know if the data of a given distribution are heavy-tailed or light-tailed when compared to the Gaussian distribution. Thus, to calculate the value of the kurtosis coefficient, we will consider three steps.

Firstly, for a variable y following the Gaussian distribution $N(y)$, we will define the mean of y as follows:

$$\mu = \int_{-\infty}^{+\infty} N(y) y\, dy \, , \qquad \text{(C.164)}$$

it is also known as the first order moment $E[y]$ of the Gaussian distribution.

Secondly, we will define the standardized kurtosis coefficient as the standardized fourth order moment,

$$K = \frac{E[(x-\mu)^4]}{(E[(x-\mu)^2])^2}; \qquad (C.165)$$

where $E[(x-\mu)^2]$ is the second order moment of the Gaussian distribution. The previously defined kurtosis coefficient can be also expressed this way:

$$K = E\left[\left(\frac{x-\mu}{\sigma}\right)^4\right], \qquad (C.166)$$

where we have

$$E\left[\left(\frac{x-\mu}{\sigma}\right)^4\right] = \int_{-\infty}^{+\infty} N(x)\left(\frac{X-\mu}{\sigma}\right)^4 dx\,. \quad (C.167)$$

Thirdly and finally, we will find out the value of the kurtosis coefficient by using the last expression, that is, $E\left[\left(\frac{x-\mu}{\sigma}\right)^4\right]$. Thereupon, in Eq. (C.167), we will replace the value of the gaussian distribution $N(x)$,

$$E\left[\left(\frac{X-\mu}{\sigma}\right)^4\right] = \int_{-\infty}^{+\infty} \frac{1}{\sqrt{2\pi}\sigma} \exp\left[-\frac{1}{2}\left(\frac{x-\mu}{\sigma}\right)^2\right]\left(\frac{x-\mu}{\sigma}\right)^4 dx\,; \quad (C.168)$$

after doing the change of variable $t = \frac{x-\mu}{\sqrt{2}\sigma}$, we have $dt = \frac{dx}{\sqrt{2}\sigma}$. Therefore, the kurtosis coefficient is given by

$$E\left[\left(\frac{X-\mu}{\sigma}\right)^4\right] = \frac{1}{\sqrt{2\pi}\sigma}\int_{-\infty}^{+\infty} \exp[-t^2]\left(t\sqrt{2}\right)^4 \sqrt{2}\sigma\, dt\,; \qquad (C.169)$$

simplifying, it follows that

$$E\left[\left(\frac{X-\mu}{\sigma}\right)^4\right] = \frac{(\sqrt{2})^4}{\sqrt{\pi}}\int_{-\infty}^{+\infty} \exp[-t^2]\, t^4\, dt = E\left[\left(\frac{X-\mu}{\sigma}\right)^4\right] = \frac{4}{\sqrt{\pi}}\int_{-\infty}^{+\infty} \exp[-t^2]\, t^4\, dt\,; \ (C.170)$$

now, we will use the following formula:

$$\int_{-\infty}^{+\infty} \exp\left[-\frac{t^2}{2}\right] t^4\, dt = 2\int_{0}^{+\infty} \exp\left[-\frac{t^2}{2}\right] t^4\, dt = 2x\frac{3!!}{8}\sqrt{\pi} = \frac{3}{4}\sqrt{\pi}\,, \ (C.171)$$

where the symbol !! stands for the double factorial, which is a generalized way of the factorial of an integer number. Thereupon, it is known that $3!! = 3$. Then, we replace the result of (C.171) in (C.170) and obtain

$$E\left[\left(\frac{X-\mu}{\sigma}\right)^4\right] = \frac{4}{\sqrt{\pi}}\frac{3}{4}\sqrt{\pi} = 3\,, \qquad (C.172)$$

which is a permanently accepted result for the kurtosis coefficient of the Gaussian distribution.

Demonstrating How to Obtain the Skewness for the Standard Gaussian Distribution

The concept of skewness was another breakthrough introduced by Karl Pearson in 1905. It allows us to quantify how much a given distribution deviates from the Gaussian distribution. Nowadays, the skewness is defined as the standardized third moment this way:

$$\gamma_1 = E\left[\left(\frac{X-\mu}{\sigma}\right)^3\right] \equiv \frac{E[(x-\mu)^3]}{\{E[(X-\mu)^2]\}^{3/2}}\,; \qquad \text{(C.173)}$$

next, we will use only the formula $\gamma_1 = E\left[\left(\frac{X-\mu}{\sigma}\right)^3\right]$. Thus,

$$\gamma_1 = E\left[\left(\frac{X-\mu}{\sigma}\right)^3\right] = \int_{-\infty}^{+\infty} N(x)\left(\frac{X-\mu}{\sigma}\right)^3 dx\,, \qquad \text{(C.174)}$$

by using the Gaussian distribution $N(x)$, it follows that the skewness is given by

$$\gamma_1 = \int_{-\infty}^{+\infty} \frac{1}{\sqrt{2\pi}\sigma} \exp\left[-\frac{1}{2}\left(\frac{x-\mu}{\sigma}\right)^2\right]\left(\frac{x-\mu}{\sigma}\right)^3 dx\,; \qquad \text{(C.175)}$$

now, we will make a change of variable, that is, $t = \frac{x-\mu}{\sqrt{2}\sigma}$; then, we have $dt = \frac{dx}{\sqrt{2}\sigma}$. Consequently, the skewness is expressed as

$$\gamma_1 = \frac{1}{\sqrt{2\pi}\sigma}\int_{-\infty}^{+\infty} \exp[-t^2]\left(t\sqrt{2}\right)^3 \sqrt{2}\sigma\, dt\,; \qquad \text{(C.176)}$$

simplifying this expression, we have as follows:

$$\gamma_1 = \frac{\left(\sqrt{2}\right)^3}{\sqrt{\pi}}\int_{-\infty}^{+\infty} \exp[-t^2]\, t^3\, dt\,; \qquad \text{(C.177)}$$

now, we realize we deal with an odd integrand evaluated in an even interval; then, the integral is zero. Therefore, the skewness of the Gaussian distribution is

$$\gamma_1 = 0\,, \qquad \text{(C.178)}$$

which is the expected result, because the Gaussian distribution is symmetrical about its mean μ.

Demonstrations Concerning the *q*-Gaussian Distribution

Demonstrating How to Obtain the q-Gaussian Distribution From Tsallis Entropy

For a continuous variable x, we start by setting $p(x) \equiv N(x)$; then, Tsallis entropy of a continuous variable x is defined by

$$S_q = S[N(x)] = \frac{k}{q-1}\left\{1 - \int_x [N(x)]^q\, dx\right\}, \qquad \text{(C.179)}$$

now, we will use the maximum entropy principle along with the Lagrange multipliers. Thus, firstly, we will consider two constraints: normalization condition and requirement of fixed variance, respectively,

$$\int_{-\infty}^{+\infty} N(x)\, dx = 1 \quad \text{and} \quad \sigma^2 = \int_{-\infty}^{+\infty} [N(x)]^q (x-\mu)^2\, dx\,; \qquad \text{(C.180)}$$

secondly, we set the following Lagrangian function:

$$L = S_q - \alpha\int_{-\infty}^{\infty} N(x)\, dx - \beta\int_{-\infty}^{+\infty} [N(x)]^q (x-\mu)^2\, dx\,, \qquad \text{(C.181)}$$

after replacing the value of Tsallis entropy,

$$L=\frac{k}{q-1}\left\{1-\int_x [N(x)]^q\,dx\right\}-\alpha\int_{-\infty}^{+\infty}N(x)\,dx-\beta\int_{-\infty}^{+\infty}[N(x)]^q(x-\mu)^2\,dx\ ,\text{(C.182)}$$

now, we group terms

$$L=\frac{k}{q-1}+\int_{-\infty}^{+\infty}\left[-\frac{k}{q-1}[N(x)]^q-\alpha N(x)-\beta[N(x)]^q(x-\mu)^2\right]dx\ ,\ \text{(C.183)}$$

then, we differentiate the Lagrangian function respect to $N(x)$ and equalize it to zero, that is,

$$\frac{\partial L}{\partial N}=\frac{\partial}{\partial N}\left\{\frac{k}{q-1}+\int_{-\infty}^{+\infty}\left[-\frac{k}{q-1}[N(x)]^q-\alpha N(x)-\beta[N(x)]^q(x-\mu)^2\right]dx\right\}=0\ ,\ \text{(C.184)}$$

after differentiating, it becomes thus:

$$\frac{\partial L}{\partial N}=\int_{-\infty}^{+\infty}\left[-\frac{k}{q-1}q[N(x)]^{q-1}-\alpha-\beta q[N(x)]^{q-1}(x-\mu)^2\right]dx=0\ ,\ \text{(C.185)}$$

therefore, we can establish that

$$-\frac{k}{q-1}q[N(x)]^{q-1}-\alpha-\beta q[N(x)]^{q-1}(x-\mu)^2=0\ ,\quad \text{(C.186)}$$

or, after factoring out $[N(x)]^{q-1}$,

$$\left[\frac{kq}{q-1}+\beta q(x-\mu)^2\right][N(x)]^{q-1}=-\alpha\ .\quad \text{(C.187)}$$

next, we multiply by $\frac{q-1}{kq}$ on both sides of the previous equation. Then,

$$\left[1+\frac{\beta}{k}(q-1)(x-\mu)^2\right][N(x)]^{q-1}=-\frac{q-1}{kq}\alpha\ ,\qquad \text{(C.188)}$$

solving for $[N(x)]^{q-1}$, we obtain this way:

$$[N(x)]^{q-1}=\frac{\frac{1-q}{kq}\alpha}{1-\frac{\beta}{k}(1-q)(x-\mu)^2}\ ,\qquad \text{(C.189)}$$

and raising to the power -1 on both sides,

$$[N(x)]^{1-q}=\frac{1-\frac{\beta}{k}(1-q)(x-\mu)^2}{\frac{1-q}{kq}\alpha}\ ,\qquad \text{(C.190)}$$

likewise, we raise it to the power $\frac{1}{1-q}$,

$$N(x)=\frac{1}{\left[\frac{1-q}{kq}\alpha\right]^{\frac{1}{1-q}}}\left[1-\frac{\beta}{k}(1-q)(x-\mu)^2\right]^{\frac{1}{1-q}}\ ,\qquad \text{(C.191)}$$

in order to find a value for $\left[\frac{1-q}{kq}\alpha\right]^{\frac{1}{1-q}}$, we integrate from $-\infty$ to $+\infty$

$$\int_{-\infty}^{+\infty}N(x)\,dx=1=\frac{1}{\left[\frac{1-q}{kq}\alpha\right]^{\frac{1}{1-q}}}\int_{-\infty}^{+\infty}\left[1-\frac{\beta}{k}(1-q)(x-\mu)^2\right]^{\frac{1}{1-q}}dx\ ,\ \text{C.192)}$$

thus, it follows that

$$\left[\frac{1-q}{kq}\alpha\right]^{\frac{1}{1-q}}=\int_{-\infty}^{+\infty}\left[1-\frac{\beta}{k}(1-q)(x-\mu)^2\right]^{\frac{1}{1-q}}dx\ ,\ \text{(C.193)}$$

by replacing Eq. (C.193) into Eq. (C.191), the probability distribution is given by

$$N(x) = \frac{\left[1-\frac{\beta}{k}(1-q)(x-\mu)^2\right]^{\frac{1}{1-q}}}{\int_{-\infty}^{+\infty}\left[1-\frac{\beta}{k}(1-q)(x-\mu)^2\right]^{\frac{1}{1-q}}dx}, \quad (C.194)$$

after applying the commutative property to $\frac{\beta}{k}(1-q)$, we obtain this way:

$$N(x) = \frac{\left[1-(1-q)\frac{\beta}{k}(x-\mu)^2\right]^{\frac{1}{1-q}}}{\int_{-\infty}^{+\infty}\left[1-(1-q)\frac{\beta}{k}(x-\mu)^2\right]^{\frac{1}{1-q}}dx}; \quad (C.195)$$

then, we recognize two q-exponentials in the last equation. Thus, we have arrived at the q-gaussian distribution

$$N(x) = \frac{\exp_q\left[\frac{\beta}{k}(x-\mu)^2\right]}{\int_{-\infty}^{+\infty}\exp_q\left[\frac{\beta}{k}(x-\mu)^2\right]dx}, \quad (C.196)$$

or

$$N(x) = \frac{1}{A_q}\exp_q\left[\frac{\beta}{k}(x-\mu)^2\right], \quad (C.197)$$

where A_q is the normalization parameter defined in the following way: $A_q = \int_{-\infty}^{+\infty}\exp_q\left[\frac{\beta}{k}(x-\mu)^2\right]dx$. Henceforth, to simplify calculations, we will consider $k = 1$.

Demonstrating How to Obtain the Second Order q-Moment of the q-Gaussian Distribution

In a general way, the second order q-moment is defined as

$$E_q\left[\left(\frac{x-\mu}{\sigma}\right)^2\right] = \int_{-\infty}^{+\infty} N(x)\left(\frac{x-\mu}{\sigma}\right)^2 dx\,; \quad (C.198)$$

henceforth, we will consider the mean $\mu = 0$ and the standard deviation $\sigma = 1$. Then, $E_q\left[\left(\frac{x-\mu}{\sigma}\right)^2\right] = E_q[x^2]$; therefore,

$$E_q[x^2] = \int_{-\infty}^{+\infty} N(x)x^2\,dx\,; \quad (C.199)$$

where $N(x)$ is the q-Gaussian distribution given by

$$N(x) = \frac{1}{C_q}\exp_q(-\beta x^2), \quad (C.200)$$

being $\beta = \frac{1}{3-q}$. Also, C_q is the normalization constant defined in this manner:

$$C_q = \frac{1}{\sqrt{\beta}}\int_{-\infty}^{+\infty}\exp_q(-x^2)\,dx = \begin{cases} \frac{1}{\sqrt{\beta}}\left(\frac{1}{q-1}\right)^{1/2} B\left(\frac{3-q}{2q-2},\frac{1}{2}\right), & 1 < q < 3 \\ \frac{1}{\sqrt{\beta}}\sqrt{\pi}, & q = 1 \\ \frac{1}{\sqrt{\beta}}\left(\frac{1}{1-q}\right) B\left(\frac{2-q}{1-q},\frac{1}{2}\right)^{1/2} B\left(\frac{2-q}{1-q},\frac{1}{2}\right), & -\infty < q < 1 \end{cases} \quad (C.201)$$

where we have the Beta functions $B\left(\frac{3-q}{2q-2},\frac{1}{2}\right)$ and $B\left(\frac{2-q}{1-q},\frac{1}{2}\right)$. After considering $1 \leq q < 3$, we have $C_q = \frac{1}{\sqrt{\beta}}\left(\frac{1}{q-1}\right)^{1/2} B\left(\frac{3-q}{2q-2},\frac{1}{2}\right)$. Thus, the second order q-moment is given by

$$E_q[x^2] = \frac{1}{C_q}\int_{-\infty}^{+\infty} x^2[1-(1-q)\beta x^2]^{\frac{1}{1-q}}\, dx\,, \qquad \text{(C.202)}$$

now, we will use the method of integration by parts, that is, we will do as follows:

$$E_q[x^2] = \frac{1}{C_q}\int_{-\infty}^{+\infty} u dv = \frac{1}{C_q}\left\{\int_{-\infty}^{+\infty} d(uv) - \int_{-\infty}^{+\infty} v du\right\}, \qquad \text{(C.203)}$$

where $u = x$ and $dv = x[1-(1-q)\beta x^2]^{\frac{1}{1-q}}dx$; then, we have $du = dx$ and $v = \int x[1-(1-q)\beta x^2]^{\frac{1}{1-q}}dx$. For this undefined integral, we will make the following change of variable: $z = 1-(1-q)\beta x^2$, then $dz = -(1-q)\beta 2x dx$. Therefore,

$$v = \int x[1-(1-q)\beta x^2]^{\frac{1}{1-q}}dx = \frac{1}{-(1-q)2\beta}\int z^{\frac{1}{1-q}}dz\,, \quad \text{(C.204)}$$

so,

$$v = \frac{-1}{(1-q)2\beta}\frac{z^{\frac{1}{1-q}+1}}{\frac{1}{1-q}+1} = \frac{-1}{2\beta(2-q)} z^{\frac{2-q}{1-q}}\,; \qquad \text{(C.205)}$$

or, as a function of x,

$$v = \frac{-1}{2\beta(2-q)}[1-(1-q)\beta x^2]^{\frac{2-q}{1-q}}\,; \qquad \text{(C.206)}$$

thus, the second order q-moment is expressed as

$$E_q[x^2] = \frac{1}{C_q}\left\{\int_{-\infty}^{+\infty} d(uv) - \int_{-\infty}^{+\infty} v du\right\}, \qquad \text{(C.207)}$$

the integral $\int_{-\infty}^{+\infty} d(uv)$ is zero, then considering $v = \frac{-1}{2\beta(2-q)} z^{\frac{2-q}{1-q}}$ and $du = dx$, we have

$$E_q[x^2] = \frac{-1}{C_q}\int_{-\infty}^{+\infty} v du = \frac{-1}{C_q}\int_{-\infty}^{+\infty}\frac{-1}{2\beta(2-q)}[1-(1-q)\beta x^2]^{\frac{2-q}{1-q}}dx\,, \text{(C.208)}$$

therefore

$$E_q[x^2] = \frac{-1}{2\beta(2-q)C_q}\int_{-\infty}^{+\infty}[1-(1-q)\beta x^2]^{\frac{2-q}{1-q}}dx\,; \quad \text{(C.209)}$$

now, in the integrand, we recognize the q-exponential function raised to the power $(2-q)$, then

$$E_q[x^2] = \frac{1}{2\beta(2-q)C_q}\int_{-\infty}^{+\infty}\left[\exp_q(-\beta x^2)\right]^{2-q} dx\,; \qquad \text{(C.210)}$$

it is time to do a math trick. Specifically, we will search for expressing $\left[\exp_q(-\beta x^2)\right]^{2-q}$ as a new power-free q-exponential: $\left[\exp_q(-\beta x^2)\right]^{2-q} = \exp_{q_1}(-\beta_1 x^2)$, where we have introduced the new parameters q_1 and β_1. Next, to find out the values of q_1 and β_1, we will consider the definition of the q-exponential function:

$$\left[\exp_q(-\beta x^2)\right]^{2-q} = [1-(1-q)\beta x^2]^{\frac{2-q}{1-q}}, \quad \text{(C.211)}$$

now, we propose $\frac{2-q}{1-q} = \frac{1}{1-q_1}$; then, after solving for q_1 and $1-q$, we have as follows: $q_1 = \frac{1}{2-q}$ (with $1 \le q_1 < 3$), and $1-q = (2-q)(1-q_1)$.

Next, we will search an expression for $(1-q)\beta$ in order to replace it into the previous numbered equation. Thus,

$$(1-q)\beta = (2-q)(1-q_1)\beta = (1-q_1)\beta_1, \quad \text{(C.212)}$$

where $\beta_1 = (2-q)\beta$. Consequently, we have the expected result:

$$\left[\exp_q(-\beta x^2)\right]^{2-q} = [1-(1-q_1)\beta_1 x^2]^{\frac{1}{1-q_1}}, \quad \text{(C.213)}$$

or

$$\left[\exp_q(-\beta x^2)\right]^{2-q} = \exp_{q_1}(-\beta_1 x^2). \quad \text{(C.214)}$$

Then, the second order q-moment is given by

$$E_q[x^2] = \frac{1}{2\beta(2-q)\mathrm{C}_q}\int_{-\infty}^{+\infty}\left[\exp_q(-\beta x^2)\right]^{2-q} dx =$$

$$= \frac{1}{2\beta(2-q)\mathrm{C}_q}\int_{-\infty}^{+\infty}\exp_{q_1}(-\beta_1 x^2)\ dx, \quad \text{(C.215)}$$

now, we will define a new normalization constant,

$$\mathrm{C}_{q_1} = \frac{1}{\sqrt{\beta_1}}\int_{-\infty}^{+\infty}\exp_{q_1}(-x^2)\ dx \equiv \int_{-\infty}^{+\infty}\exp_{q_1}(-\beta_1 x^2)\ dx; \quad \text{(C.216)}$$

thus, we arrive at

$$E_q[x^2] = \frac{1}{2\beta(2-q)}\frac{\mathrm{C}_{q_1}}{\mathrm{C}_q}. \quad \text{(C.217)}$$

Next, we will replace the constant values C_{q_1} and C_q and calculate the $\frac{\mathrm{C}_{q_1}}{\mathrm{C}_q}$ ratio. Then,

$$\frac{\mathrm{C}_{q_1}}{\mathrm{C}_q} = \frac{\frac{1}{\sqrt{\beta_1}}\left(\frac{1}{q_1-1}\right)^{1/2} B\left(\frac{3-q_1}{2q_1-2},\frac{1}{2}\right)}{\frac{1}{\sqrt{\beta}}\left(\frac{1}{q-1}\right)^{1/2} B\left(\frac{3-q}{2q-2},\frac{1}{2}\right)}, \quad \text{(C.218)}$$

by replacing $q_1 = \frac{1}{2-q}$ and $\beta_1 = (2-q)\beta$, it follows that

$$\frac{\mathrm{C}_{q_1}}{\mathrm{C}_q} = \frac{\frac{1}{\sqrt{(2-q)\beta}}\left(\frac{1}{\frac{1}{2-q}-1}\right)^{1/2} B\left(\frac{3-\frac{1}{2-q}}{2\frac{1}{2-q}-2},\frac{1}{2}\right)}{\frac{1}{\sqrt{\beta}}\left(\frac{1}{q-1}\right)^{1/2} B\left(\frac{3-q}{2q-2},\frac{1}{2}\right)}, \quad \text{(C.219)}$$

simplifying

$$\frac{\mathrm{C}_{q_1}}{\mathrm{C}_q} = \frac{B\left(\frac{5-3q}{2q-2},\frac{1}{2}\right)}{B\left(\frac{3-q}{2q-2},\frac{1}{2}\right)}, \quad \text{(C.220)}$$

now, we realize $\frac{5-3q}{2q-2}+1=\frac{3-q}{2q-2}$; then, we will use a recurrence formula for the Beta function,

$$B(a+1,b)=\frac{a}{a+b}B(a,b)\,, \qquad \text{(C.221)}$$

with $a=\frac{5-3q}{2q-2}$, $a+1=\frac{3-q}{2q-2}$, and $b=\frac{1}{2}$. Specifically, we will manipulate $B\left(\frac{3-q}{2q-2},\frac{1}{2}\right)$,

$$B\left(\frac{3-q}{2q-2},\frac{1}{2}\right)=\frac{\frac{5-3q}{2q-2}}{\frac{5-3q}{2q-2}+\frac{1}{2}}B\left(\frac{5-3q}{2q-2},\frac{1}{2}\right)=\frac{5-3q}{4-2q}B\left(\frac{5-3q}{2q-2},\frac{1}{2}\right); \qquad \text{(C.222)}$$

so, Eq. (C.220) is given by

$$\frac{C_{q_1}}{C_q}=\frac{B\left(\frac{5-3q}{2q-2},\frac{1}{2}\right)}{\frac{5-3q}{4-2q}B\left(\frac{5-3q}{2q-2},\frac{1}{2}\right)}\,, \qquad \text{(C.223)}$$

after simplifying, we get

$$\frac{C_{q_1}}{C_q}=\frac{2(2-q)}{5-3q}\;; \qquad \text{(C.224)}$$

therefore, in accordance with Eq. (C.217), the second order q-moment is expressed as

$$E_q[x^2]=\frac{1}{\beta(5-3q)}\,; \qquad \text{(C.225)}$$

finally, we replace the value of $\beta=\frac{1}{3-q}$ and obtain thus:

$$E_q[x^2]=\frac{3-q}{5-3q}\,. \qquad \text{(C.226)}$$

Demonstrating How to Obtain the Kurtosis Coefficient for the q-Gaussian Distribution

We start by remembering that the kurtosis coefficient is defined as the standardized fourth-order q-moment, that is,

$$K=E_q\left[\left(\frac{x-\mu}{\sigma}\right)^4\right]; \qquad \text{(C.227)}$$

henceforth, we will use $K=E_q[x^4]$, that is, we will consider $\mu=0$ and $\sigma=1$. Then,

$$K=E_q[x^4]=\int_{-\infty}^{+\infty}N(x)x^4\,dx\,, \qquad \text{(C.228)}$$

after replacing the value of the q-Gaussian distribution $N(x)=\frac{1}{C_q}\exp_q(-\beta x^2)$, the kurtosis coefficient is given by

$$K=\frac{1}{C_q}\int_{-\infty}^{+\infty}[1-(1-q)\beta x^2]^{\frac{1}{1-q}}x^4\,dx\,, \qquad \text{(C.229)}$$

now, we will use the method of integration by parts with $u=x^3$ and $dv=x[1-(1-q)\beta x^2]^{\frac{1}{1-q}}dx$; then, we have $du=3x^2dx$ and $v=$

$\int x[1-(1-q)\beta x^2]^{\frac{1}{1-q}}dx$. Also, we already know the result for the last integral: $v = \frac{-1}{2\beta(2-q)}[1-(1-q)\beta x^2]^{\frac{2-q}{1-q}}$. Thus, we have

$$K = \frac{1}{2\beta(2-q)\mathrm{C}_q}\int_{-\infty}^{+\infty}[1-(1-q)\beta x^2]^{\frac{2-q}{1-q}}\,3x^2dx\,; \qquad \text{(C.230)}$$

furthermore, we have already demonstrated $[1-(1-q)\beta x^2]^{\frac{2-q}{1-q}} = \exp_{q_1}(-\beta_1 x^2)$, where we have introduced the new parameters q_1 and β_1, with $q_1 = \frac{1}{2-q}$ and $\beta_1 = (2-q)\beta$. Then, the kurtosis coefficient of the q-Gaussian distribution is given by

$$K = \frac{3}{2\beta(2-q)\mathrm{C}_q}\int_{-\infty}^{+\infty}\exp_{q_1}(-\beta_1 x^2)\,x^2dx\,; \qquad \text{(C.231)}$$

we realize the last integral is $\mathrm{C}_{q_1}E_{q_1}[x^2]$, therefore,

$$K = \frac{3}{2\beta(2-q)}\frac{\mathrm{C}_{q_1}}{\mathrm{C}_q}E_{q_1}[x^2]\,; \qquad \text{(C.232)}$$

now, from Eq. (C.224), we remember about $\frac{\mathrm{C}_{q_1}}{\mathrm{C}_q} = \frac{2(2-q)}{5-3q}$; besides, we also know that $\beta = \frac{1}{3-q}$. Likewise, from Eq. (C.226), we can infer the formula $E_{q_1}[x^2] = \frac{3-q_1}{5-3q_1}$. Therefore, by replacing these expressions in Eq. (C.232), it follows that

$$K = \frac{3}{2\frac{1}{3-q}(2-q)}\frac{2(2-q)}{5-3q}\frac{3-q_1}{5-3q_1}\,, \qquad \text{(C.233)}$$

and by using $q_1 = \frac{1}{2-q}$, K is expressed as

$$K = \frac{3(3-q)}{5-3q}\frac{(5-3q)}{(7-5q)}\,; \qquad \text{(C.234)}$$

after cross canceling $(5-3q)$, the final expression for the kurtosis coefficient of the q-Gaussian distribution is given by

$$K = \frac{3(3-q)}{7-5q}\,. \qquad \text{(C.235)}$$

Bibliography

Beck, Christian. 2005. "Superstatistics: Recent developments and applications." In *Complexity, Metastability and Nonextensivity*, The Science and Culture Series — Physics, edited by C. Beck, G. Benedek, A. Rapisarda, and C. Tsallis. 33–42. Singapore: World Scientific.

Beck, Christian, and Cohen, Ezechiel. 2003. "Superstatistics," *Phys. A*, Vol. 322, Iss. C, 267–275.

Busby, Chris, and Howard, Charles. 2017. "Re-Thinking Biology—I. Maxwell's Demon and the Spontaneous Origin of Life," *Adv. Biol. Chem.*, Vol. 7, 170–181.

Dagotto, Elbio, Hotta, Takashi, and Moreo, Adriana. 2001. "Colossal magnetoresistant materials: the key role of phase separation," *Phys. Rep.*, Vol. 344, Iss. 1–3, p. 1–153.

Dagotto, Elbio. 2005. "Open questions in CMR manganites, relevance of clustered states and analogies with other compounds including the cuprates," *New J. Phys.*, Vol. 7, 67, 1–28.

Deac, Iosif, Mitchell, John, and Schiffer, Peter. 2001. "Phase separation and low-field bulk magnetic properties of $Pr_{0.7}Ca_{0.3}MnO_3$," *Phys. Rev. B*, Vol. 63, Iss. 17, 172408.

Flamm, Dieter. 1997. "History and outlook of statistical physics," IN: Conference on Creativity in Physics Education, August 23, 1997, Sopron, Hungary (UWThPh-1997-52).

Gschneidner, Karl, and Pecharsky, Vitalij. 2000. "Magnetocaloric materials," *Annu. Rev. Mater. Sci.*, Vol. 30, 387–429.

Haghiri-Gosnet, Anne-Marie, and Renard, Jean-Pierre. 2003. "CMR manganites: Physics, thin films and devices," *J. Phys. D: Appl. Phys.*, Vol. 36, R127.

Hardy, Vincent, Wahl, Alexandre, Martin, Christine, and Simon, Charles. 2001. "Low-temperature specific heat in $Pr_{0.63}Ca_{0.37}MnO_3$: Phase separation and metamagnetic transition," *Phys. Rev. B,* Vol. 63, Iss. 22, 224403.

Jirak, Zdenek, Krupicka, Svatopluk, Simsa, Zdeněk, Dlouha, Mája, and Vratislav, Stanislav. 1985. "Neutron diffraction study of $Pr_{1-x}Ca_xMnO_3$ perovskites," *J. Magn. Magn. Mater.*, Vol. 53,153–166.

Kaganov, Moisei, and Lifshits, Il'ya. 1979. "Electron theory of metals and geometry," *Sov. Phys. Uspekhi*, Vol. 22, Iss. 11, 904.

Khomskii, Daniel. 2001. "Orbital effects in manganites," *Int. J. Mod. Phys. B*, Vol. 15, No. 19–20, 2665–2681.

Lea, Kenneth, Leask, Michael, and Wolf, Werner. 1962. "The raising of angular momentum degeneracy of f-electron terms by cubic crystal fields," *J. Phys. Chem. Solids*, Vol. 23, Iss. 10, 1381–1405.

Nagaev, E. 1996. "Magnetoimpurity theory of resistivity and magnetoresistance for degenerate ferromagnetic semiconductors of the $LaMnO_3$ type," *Phys. Rev. B*, Vol. 54, Iss. 23, 16608–16617.

Pecharsky, Vitalij, and Gschneidner, Karl. 1999. "Magnetocaloric effect from indirect measurements: Magnetization and heat capacity," *J. App. Phys.*, Vol. 86, 565.

Reis Jr., Mario de Souza. 2003. "*Não-Extensividade magnética em manganitas*" (Magnetic nonextensivity in manganites), Doctoral thesis, Brazilian Center for Research in Physics (CBPF), Rio de Janeiro, Brazil.

Rizzo, Salvo, and Rapisarda, Andrea. 2005. "Application of superstatistics to atmospheric turbulence," IN: Complexity, Metastability and Nonextensivity, pp. 246–252, edited by C. Tsallis, A. Rapisarda, G. Benedek, and C. Beck

Satou, Atsuo, and Yamanaka, Masanori. 2001. "Cantor spectra for the double-exchange model," *Phys. Rev. B*, Vol. 63, 212403.

Tishin, Alexander, Spichkin, Yu, Zvereva, Vladimir, and Egolfd, Peter. 2016. "A review and new perspectives for the magnetocaloric effect: New materials and local heating and cooling inside the human body," *Int. J. Refrigeration.*, Vol. 68, 177–186.

Tsallis, Constantino. 1988. "Possible generalization of Boltzmann-Gibbs statistics," *J. Stat. Phys.,* Vol. 52, 479–487.

van den Brink, Jeroen, Khaliullin, Giniyat, and Khomskii, Daniel. 2004. "Orbital effects in manganites," pp. 263–301. IN: Colossal Magnetoresistive Manganites, edited by T. Chatterji, Kluwer Academic Publishers.

Website 1. 2022. http://tsallis.cat.cbpf.br/TEMUCO.pdf (accessed May 24, 2022).

Website 2. 2022. https://commons.wikimedia.org/wiki/File:Perovskite_ABO3.jpg (accessed May 24, 2022).

INDEX

algorithm
    DM Renormalization Group, 27
Bayesian probability, 22
bra, 79
Cantor set, 11
CE-phase, 11
constraints, 43
coupling
    interlattice, 74
    intra-lattice, 74
cubic perovskite, 5, 6
degrees of freedom, 19
dipole moment, 41
dry etching, 18
effect
    Hall, 100
    Jahn-Teller, 6
    magnetocaloric, 18, 102
    spin-Hall, 101
electrical conductivity, 11
energy parameter, 44
entropic index, 38
entropy
    Boltzmann, 28
    informational, 25
    optimization, 44
    quantum group, 29
    Renyi, 37
    Shannon, 29
    thermodynamic, 28
    Tsallis, 32
    κ-, 30
escort
    distributions, 42
    entropy, 43
Griffiths singularities, 14
Heussler alloys, 103
historical events, 22
Hund's rule, 8
inherent defects, 13
irreversibility, 16
Jahn-Teller
    ion, 6
    phonon, 15
    polarons, 11
ket, 79
kinks, 70
klitzing, 100
Lagrange multiplier, 32, 39, 44
law of thermodynamics
    second, 29
Lévy distributions, 25
macrostate, 28
magnetoresistance
    CMR, 4
    giant, 101
manganese oxides, 1
manganites, 1
    mother compound of, 5
    nanostructured, 17
    perovskites, 4, 5
    pyrochlore, 7
Maxwell's demon, 44
mean-field approximation, 74
metal-insulator transition, 15
microstate, 28
model
    Anderson localization, 11
    double-exchange, 4, 10
    electron gas, 16
    Ising, 25
    $J_{H=\infty}$, 17
    Oguchi, 94
    one-orbital, 12, 16
    phase separation scenario, 12
    Phononic Jahn-Teller, 17
    polarons, 11
    two-orbital, 12

molecular-field approximation, 73
nonextensive
    internal energy, 42, 43, 71
    magnetization, 41, 44, 71
    quasi-probability distributions, 1
    statistical mechanics, 1
ordering
    charge, 14
    orbital
        e_g, 15
partition function, 44
phase
    Griffiths, 13
phases
    competition and coexistence, 12, 13
polytropes, 33
principle
    Jaynes, 26, 44
    Kullback, 44
prototype material, 13
pseudo-additivity, 39
renormalization group, 27
self-similarity, 11
spin-glass, 13
spintronics, 102
SQUID, 87
statistical
    ensemble, 24
statistical independence, 38
statistical mechanics
    Boltzmann-Gibbs-Shannon, 1
Stevens operators, 95, 96
superexchange, 10
superstatistics, 30
systems
    compound, 1
    statistically independent, 38
transition
    Verwey, 14
tunneling devices, 19